Molecular Biotechnology

Molecular

S.B. PRIMROSE

BSc, PhD
Amersham International plc
Amersham
Buckinghamshire

Biotechnology

SECOND EDITION

OXFORD

Blackwell Scientific Publications

LONDON EDINBURGH BOSTON

MELBOURNE PARIS BERLIN VIENNA

© 1987, 1991 by
Blackwell Scientific Publications
Editorial offices:
Osney Mead, Oxford OX2 0EL
25 John Street, London WC1N 2BL
23 Ainslie Place, Edinburgh EH3 6AJ
3 Cambridge Center, Cambridge
  Massachusetts 02142, USA
54 University Street, Carlton
  Victoria 3053, Australia

Other Editorial Offices:

Arnette SA
2, rue Casimir-Delavigne
75006 Paris
France

Blackwell Wissenschaft
Meinekestrasse 4
D-1000 Berlin 15
Germany

Blackwell MZV
Feldgasse 13
A-1238 Wien
Austria

First published under the title *Modern Biotechnology* 1987
Reprinted 1989
Second edition 1991

Set by Setrite Typesetters, Hong Kong
Printed and bound in Spain
by Printek s.a., Bilbao

DISTRIBUTORS
Marston Book Services Ltd
PO Box 87
Oxford OX2 0DT
(*Orders*: Tel: 0865 791155
        Fax: 0865 791927
        Telex: 837515)

USA
Blackwell Scientific Publications, Inc.
3 Cambridge Center
Cambridge, MA 02142
(*Orders*: Tel: 800 759-6102)

Canada
Oxford University Press
70 Wynford Drive
Don Mills
Ontario M3C 1J9
(*Orders*: Tel: 416 441-2941)

Australia
Blackwell Scientific Publications
(Australia) Pty Ltd
54 University Street
Carlton, Victoria 3053
(*Orders*: Tel: 03 347-0300)

British Library
Cataloguing in Publication Data

Primrose, S.B.
  Molecular biotechnology.
  1. Biotechnology
  I. Title
  660.68

  ISBN 0-632-03233-2
  ISBN 0-632-03053-4 pbk

Library of Congress
Cataloging in Publication Data

Primrose, S.B.
  Molecular biotechnology/S.B. Primrose. — 2nd ed.
      p.   cm.
  Rev. ed. of: Modern biotechnology. 1987.
  Includes bibliographical references and index.
  ISBN 0-632-03053-4 (pb)
  ISBN 0-632-03233-2 (hb)
  1. Biotechnology. I. Primrose, S.B.
  Modern biotechnology.
  II. Title.
  [DNLM: 1. Biotechnology.   TP 248.2 P953m]
  TP248.2.P75   1991
  660'.6—dc20
  DNLM/DLC
  for Library of Congress   91−11513
                                                      CIP

Contents

Preface

Biotechnology has been one of the buzz words of the last decade. Rightly or wrongly it is often associated with easy money and, consequently, means all things to all people. So what is biotechnology? At its simplest it is the commercial exploitation of living organisms or their components, e.g. enzymes. Thus the science formerly known as industrial microbiology clearly falls within the definition of biotechnology and to many they are synonymous. But plants and animals are also exploited commercially in the practices known as horticulture and agriculture and these could come under the umbrella of biotechnology. By common consent the terms plant and animal biotechnology are restricted to that grey area encompassing the application of modern molecular biology and cell culture techniques to the manipulation of plants and animals, e.g. plant cell culture and transgenic animals, and which are not yet part of conventional plant and animal breeding.

This broad view of what constitutes biotechnology forms the basis of this book. The overriding aim has been to combine in a single text detailed information on recombinant DNA technology, protein engineering, industrial microbiology, monoclonal antibodies, plant and animal cell culture, new methods of plant and animal breeding and the legal, social and ethical issues which surround biotechnology. It provides a basic reference source for undergraduates taking courses in biochemistry, genetics and molecular biology and also for industrial specialists who seek further information on areas with which they are less familiar. The book is heavily biased towards the impact of *in vitro* gene manipulation but then without gene manipulation biotechnology would not be a buzz word!

Changes since the first edition include new sections on the many uses of the polymerase chain reaction (PCR), single domain antibodies (dABs), catalytic antibodies (abzymes), adoptive immunotherapy, gene targeting in animals and animal equivalents of human diseases. Topics which have been extensively updated include those on increasing protein stability by protein engineering, the development of agronomically useful plants by gene manipulation and the chapter on the legal, social and ethical implications of biotechnology. Key references, many of them to papers published in the last 2 years, are now cited in the text.

SANDY PRIMROSE
March 1991

Acknowledgements

Few authors can produce the finished manuscript for a textbook entirely unaided and this author is no exception. Certainly, without a secretary to bring order to chaos, to prevent interruptions at times of creativity, and to produce the typescript, this book would never have been published. For this edition, these services were willingly provided by Margaret Courtney to whom I am indebted. Thanks also are due to my wife and family for going on holiday at the relevant time to enable me to write intensively.

Abbreviations

AAT	α_1-antitrypsin
ADA	adenosine deaminase
AIDS	acquired immune deficiency syndrome
BPV	bovine papillomavirus
BST	bovine somatotropin
CaMV	cauliflower mosaic virus
CDR	complementarity-determining region
dAB	single domain antibody
DHFR	dihydrofolate reductase
ELISA	enzyme-linked immunosorbent assay
EPO	European Patent Office
EPo	erythropoietin
Fab	antigen-binding fragment
Fc	constant-region fragment
FMDV	foot and mouth disease virus
GMP	good manufacturing practice
GVH	graft-versus-host
HAT	hypoxanthine phosphoribosyl transferase
HIV	human immunodeficiency virus
HLA	human leucocyte antigen
Ig	immunoglobulin
IL	interleukin
ISH	*in situ* hybridization
LAK	lymphokine-activated killer
MAB	monoclonal antibody
MMT	mouse metallothionein
NIH	National Institutes of Health
NMR	nuclear magnetic resonance
PCR	polymerase chain reaction
RAIT	radioimmunotherapy
rbs	ribosome binding site
RFLP	restriction fragment length polymorphism
RIA	radioimmunoassay
SCID	severe combined immunodeficiency disease
SCP	single-cell protein
TIL	tumour-infiltrating lymphocyte
TK	thymidine kinase
tPA	tissue plasminogen activator
VSV	vesicular stomatitis virus

Part I
Introduction

1 Biotechnology — Ancient and Modern

Biotechnology is not a recent development. Micro-organisms have been used to produce food such as beer, vinegar, yoghurt and cheese for over eight millennia. Certainly the ancient Sumerians were familiar with beer and alehouses were an established part of Roman civilization. Wine also was popular with the Romans and they tried introducing grapevines into southern Britain for the express purpose of wine-making. References to wine and vinegar (Fig. 1.1) are scattered throughout the Bible, which is an indication that their production dates back to early times.

In many instances microbial contamination of food results in spoilage, although what is unpalatable to one person may be a delicacy to another! Occasionally microbial growth would result in beneficial changes such as improved flavour and texture and, more importantly, improved storage quality. Once these desirable changes had occurred, they would be self-perpetuating. In the absence of a knowledge of micro-biology, storage vessels would not be cleaned and the residual food would act as an inoculum. In many respects modern production of fermented foods is little different: open vessels are still used and a residue from one batch is used to inoculate the next one.

Ethanol was the first chemical to be produced with the aid of biotechnology. The origins of distillation are not clear but by the fourteenth century AD it was widely used to increase the alcoholic content of wines and beers. Indeed it was at this time when the 'auld alliance' between France and Scotland was in its prime that the French brandy manufacturers taught the Scottish brewers to distil their beer to produce whisky! From the production of spirit beverages it was but a small step to the production of neat alcohol and approximately 25% of world ethanol production is still produced by this biological route.

Until just over a century ago it was not realized that microorganisms were involved in the production of alcohol and vinegar. The discovery came when a group of French merchants were searching for a method that would prevent wine and beer from souring when they were shipped over long distances. They asked Louis

> 20 And Noah began to be an husbandman, and he planted a vineyard:
> 21 And he drank of the wine, and was drunken; and he was uncovered within his tent.

Book of Genesis, Chapter 9.

> 46 And about the ninth hour Jesus cried with a loud voice, saying, Eli, Eli, lama sabachthani? that is to say, My God, my God, why hast thou forsaken me?
> 47 Some of them that stood there, when they heard that, said, This man calleth for Elias.
> 48 And straightway one of them ran, and took a sponge, and filled it with vinegar, and put in on a reed, and gave him to drink.

The crucifixion of Christ.
Book of Matthew, Chapter 27.

Fig. 1.1 Biblical references to alcohol and vinegar.

Pasteur for help. At the time many scientists believed that air acted on the sugars in these fluids to convert them into alcohol. Instead, Pasteur found that yeasts convert the sugars to alcohol in the absence of air. Such an anaerobic process is known as *fermentation*. Souring and spoilage occur later and are due to the activities of a group of bacteria, the acetic acid bacteria, which convert alcohol into vinegar (acetic acid). Pasteur's solution was to heat the alcohol just enough to kill most of the microorganisms present, a process that does not greatly affect the flavour of the wine or beer. This process is known as *pasteurization*, although we now know that a similar technique was used for the manufacture of sake in the Orient over 300 years earlier.

Distillation apart, there was little change in bio-technology from the pre-Christian era until the early twentieth century and, as is often the case with technological advance, the impetus was provided by war. At the start of the First World War the British naval blockade prevented the Germans from importing the vegetable oils necessary to produce glycerol for explosives manufacture. Consequently the Germans turned to the microbial production of glycerol by yeast (Fig. 1.2) and soon were able to manufacture over 1000 tons per month. For its part, Germany was able to hinder the British war effort since prior to hostilities it had been the source of acetone and butanol: the former was required for munitions and the latter for

artificial rubber. The result was the British development of the acetone−butanol fermentation (Fig. 1.3) using *Clostridium acetobutylicum*.

The glycerol fermentation was short-lived. By contrast, the acetone−butanol fermentation survived until the early 1950s and during the Second World War some interesting innovations were made, such as the introduction of semi-continuous fermentation. Initially *Cl. acetobutylicum* was cultured in a large volume of medium until growth ceased. At this stage two-thirds of the medium was removed for extraction of acetone and butanol and the culture vessel filled up with fresh medium. With such a large inoculum of cells complete conversion of sugar to solvents occurred in only 12 h. Later it was realized that it was possible to use non-sterile medium; although the yield of solvent was lower due to the growth of contaminating bacteria, the

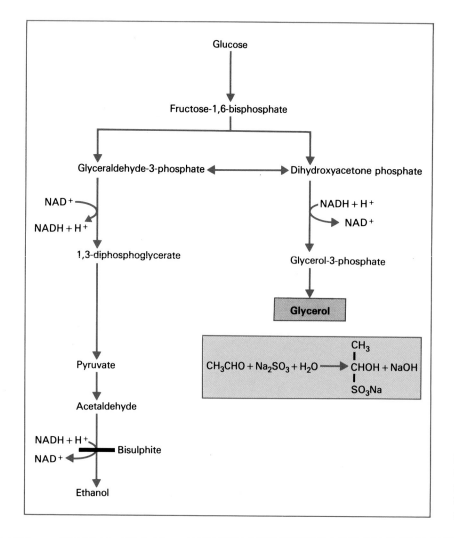

Fig. 1.2 The formation of glycerol instead of ethanol when bisulphite is added to a yeast culture. The bisulphite forms an addition product with acetaldehyde as shown in the inset.

Fig. 1.3 The pathway of acetone and butanol formation by *Clostridium acetobutylicum*. Note that hydrogen gas is produced in addition to the two solvents. Not surprisingly, a number of manufacturing plants were destroyed by explosions!

savings in fuel were substantial. In the immediate post-war period many organic chemicals, including acetone and butanol, became readily available from by-products of the petroleum industry and the fermentation process was discontinued. However, the acetone–butanol fermentation is an exceedingly simple process which might be of benefit to many Third World countries who cannot afford to spend vast sums of money on either petroleum itself or petrochemical-based products. Since these countries often have an abundance of the necessary cheap raw materials such as sugar or starch this fermentation process may make a comeback!

Present-day citric acid manufacture also has its origins in the First World War. Until then citric acid had been extracted from citrus fruits and the major producer was Italy. As men were called to arms the citrus groves were left untended. By the time hostil-

ities ended the industry was in ruins and the price of citric acid had escalated. This paved the way for the introduction of a microbial process in 1923. Unlike *Cl. acetobutylicum*, the organism used to produce citric acid (*Aspergillus niger*) is an obligate aerobe and so must be cultured in the presence of oxygen. Initially large-scale culture was achieved by placing liquid medium in shallow metal pans and allowing the organism to grow on the surface. Later this method of *surface culture* was improved by absorbing the nutrient medium on to an inert granular support (Fig. 1.4(a)).

PENICILLIN AND THE PRODUCTION OF FINE CHEMICALS [1]

Penicillin was the name given by Fleming to the antibacterial substance produced by the mould *Penicillium notatum*. By 1940 when the first purified preparations

(a)   (b)

Fig. 1.4 Production of citric acid (a) by surface culture and (b) in a stirred tank reactor. (Photos courtesy of John and E. Sturge Ltd.)

became available the phenomenal curative properties of penicillin were obvious. The significance of this discovery was not lost on a country again at war. Like the *A. niger* used to produce citric acid, *P. notatum* is an obligate aerobe and had to be grown in surface culture. Not only was this labour-intensive but the cultures were prone to contamination, thus reducing the yield of penicillin. The need for aseptic operation led to the development of the stirred tank reactor (Fig. 1.4(b)), which to this day is the preferred method for the large-scale cultivation of microbes. Aseptic operation is achieved by sterilizing all the equipment with steam prior to inoculation and by keeping the pressure inside the vessel higher than atmospheric pressure. To meet the oxygen demand of the culture, sterile air is blown into the vessel and distributed throughout the medium by agitation.

A second contribution to modern biotechnology that was made by the penicillin programme was the development of strain selection procedures. The original *P. notatum* culture yielded only 2 mg of penicillin per litre of culture fluid. By screening many different *Penicillium* isolates a higher-yielding variant, *P. chrysogenum*, was identified. In an attempt to improve the yield still further the *P. chrysogenum* was exposed systematically to a variety of mutagens such as nitrogen mustard, ultraviolet radiation and X-irradiation. After each round of exposure the survivors were screened and the highest-yielding variant carried forward for the next round of mutation. By combining fermentation improvements with the use of mutants the titre has been increased to over 20 g/l.

That microbes could produce antibiotics had been known for a long time but was considered of little significance. Once the clinical utility of penicillin was established, pharmaceutical companies began to consider antibiotics seriously. Following the discovery that *Streptomyces griseus* also produces a clinically useful antibiotic, streptomycin, it became standard practice to screen large numbers of environmental isolates for their ability to produce antibiotics. Since then it has been shown that the filamentous bacteria known as actinomycetes, a group to which *S. griseus* belongs, produce many hundreds of different antibiotics, including at least 90% of those known today. Gradually the screens became more and more sophisticated and the need to test more and more microorganisms led to the isolation of microorganisms from increasingly exotic sources. In recent times it has become increasingly difficult to find antibiotics which are both novel and useful; the number of new antibiotics discovered per year has remained constant but over the last 20 years the clinical success rate has dropped from 5% to less than 1%. Consequently pharmaceutical companies have redirected their screening efforts towards the identification of pharmacologically active fermentation products. As with the penicillin fermentation, once a microbe is identified which produces a useful metabolite, strain improvement and fermentation development are undertaken.

INDUSTRIAL USES OF PLANT AND ANIMAL CELL CULTURE

Animal cell culture has long been used in the production of viral vaccines but until the 1960s large-scale culture had not even been attempted. It was successful due largely to the vigorous application of the principles of aseptic operation formulated during the development of the penicillin fermentation. With some animal cell lines there is a problem not encountered with microbial cells, namely that the cells do not grow in suspension but require a surface for growth. To satisfy this requirement some novel solutions have been adopted. In some instances the surface area inside the vessel is increased by the addition of microcarrier beads, which are maintained in suspension; in other cases, multiple plates are fitted inside the fermenter. Despite these successes the application of mass animal cell culture has been limited largely to the production of vaccines. New avenues have opened up with the realization that certain cell lines derived from human tumours can secrete sufficiently high levels of human proteins to warrant their use for commercial production. Thus the Bowes melanoma line over-produces tissue plasminogen activator, which can be used to dissolve blood clots following coronary thrombosis. The Namalwa cell line overproduces interferons and currently is a commercial source of these antiviral and anticancer proteins. Nevertheless, the number of tumour lines identified that produce useful products is limited.

In the last 10 years certain tumour cell lines have attracted a great deal of attention and these are the *hybridomas*. As their name implies, hybridomas are hybrid cells. They are created by fusing myeloma (a type of tumour) cells with antibody-producing spleen lymphocytes. Following fusion with the myeloma cell the lymphocyte acquires immortality and can be grown indefinitely in cell culture while continuing to secrete antibody. Since any given lymphocyte only synthesizes a single antibody species, all the antibody molecules made by culture of any particular hybridoma will be identical. Since all the antibodies in the preparation are identical they are said to be *monoclonal*, i.e. they are all derived from a single clone of lymphocytes. Monoclonal antibodies can be purified easily and find application in many different areas from diagnostic kits to cancer therapy and protein purification.

For a long time the ability to cultivate plant cells in the laboratory was considered no more than a curiosity. Although whole plants serve as valuable sources of agricultural chemicals, drugs, flavourings and colourings, very few of these compounds are produced in cell culture. When they are so produced, yields are low and classical mutation and selection techniques have not been particularly successful in improving them. Today plant cell culture is of considerable importance and this is a direct result of the development of methods for the regeneration of plants from individual cells. The benefits of this are twofold. First, regeneration enables hundreds of plants to be produced in a single experiment. During regeneration genetic variants are thrown off at a high frequency and these can be very useful to the plant breeder. Second, virus-free cells can be isolated and used to produce virus-free crops, which will give increased yields per acre.

GENETIC ENGINEERING AND THE NEW BIOTECHNOLOGY

After the introduction of the penicillin fermentation there were virtually no significant new developments in industrial microbiology for 30 years. Most companies followed the practice of screening microbes for desirable metabolites or activities and then, however low the levels, implementing strain selection and fermentation development. In the late 1960s considerable excitement was generated by the prospect of using microbial cells (or *biomass*) as a source of protein, the so-called *single-cell protein* or SCP. The rationale for this was that globally there was a shortage of protein and with a rapidly expanding population this situation was going to become worse. However, the introduction of SCP has not been a success. Development coincided with a rapid increase in oil prices and the introduction of improved, high-yielding varieties of crops (Fig. 1.5). The developed countries did not need SCP, as they had a plentiful supply of protein from conventional sources; the underdeveloped countries,

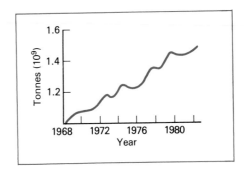

Fig. 1.5 Increase in grain production in the period 1968–1982. Similar increases in yield have been reported for other protein sources.

on the other hand, could not afford to buy SCP or even to build and run SCP plants. Thus developments in biotechnology, as in other areas, are subject to political and economic pressures. High-quality science does not guarantee commercial success and there is little altruism in the business world.

In the 1980s biotechnology became a major growth area. This change came about through a single development: the ability to splice together *in vitro* DNA molecules derived from different sources. This gene splicing ability is referred to as gene manipulation. Because one is recombining pre-existing genetic sequences to create a novel combination the term *recombinant DNA technology* is also used. Whereas mutation and selection can be used to increase the level of pre-existing activity in a microbial cell, recombinant DNA technology can be used to confer on cells entirely *new* synthetic capabilities; for example, bacteria can for the first time synthesize human hormones. This technology is not confined to microorganisms: plant and animal cells and even intact plants and animals can also be modified.

Early work on the commercial applications of recombinant DNA technology centred on the production of proteins. There were two reasons for this. Not only did the technology enable many proteins, principally of therapeutic interest, to be produced for the first time but proteins are the immediate products of a gene. As genetic engineers have become more skilled there has been a trend towards cloning all the genes associated with a biosynthetic pathway; for example, all the genes encoding the 29 steps in erythromycin biosynthesis have been cloned on a single DNA fragment. One outcome of this may be that ultimately the range of different organisms used in commercial biotechnology may diminish. Where low-molecular-weight compounds are required the pathways may be introduced to a limited selection of bacteria and fungi, e.g. *Escherichia coli*, *Streptomyces* spp., *Saccharomyces cerevisiae* and *Aspergillus* spp. Large-molecular-weight compounds may be produced in an even more restricted range of organisms and animal cells may grow in importance because they can effect post-translational modifications of proteins in an identical fashion to the intact animal (see p. 21).

Biotechnology, as practised today, is much more than recombinant DNA technology. It includes hybridoma technology, the use of cells and enzymes immobilized on inert supports and the ability to regenerate plants from isolated cells. However, if these are the per-

formers, gene manipulation is the star of the show. The ability to splice genes has revolutionized the industry and is leading to the development of countless new products and improved methods for well-established processes. The aim of this book is to put into context all the new technologies and therefore no apology is made for the heavy emphasis on recombinant DNA technology.

When the techniques of gene manipulation were first developed 15 years ago this author, like many others, anticipated a bright future for a modern biotechnology industry. However, at the time of writing (early 1991) few companies have made much money from recombinant-derived products and many fledgeling biotechnology enterprises have failed or been acquired by others (Fig. 1.6). But all this does not mean that there is no future for a 'molecular biotechnology' industry. The first fruits of the modern biotechnology industry were expected to be therapeutic proteins. Only a few of these are on the market (see p. 93) but this reflects the time taken to get regulatory approval rather than any disappointment with the products. Many other processes and products are being delayed by regulatory and political hurdles (see Chapter 14). When the first edition of this book was published in 1987 it was expected that within 1–2 years the first attempts at human gene therapy would have taken place, that recombinant bovine somatotropin would have been boosting milk yield, and that crops genetically engineered to express herbicide resistance would be growing in fields all over the Western world. That this has not occurred merely reflects an unfortunate choice of test systems. But the debate that rages is no different from that which preceded the introduction of pasteurization or oleo-margarine. In the mean time millions of US dollars, if not billions, have been made from monoclonal antibodies, which are extensively used in diagnostics (see Chapter 10).

Further reading

GENERAL REFERENCES

A general introduction to biotechnology can be found in the September 1981 issue of *Scientific American* and in the book by Steve Prentis, *Biotechnology: a New Industrial Revolution* (1984, Orbis Publishing, London). The former concentrates on industrial microbiology, whereas the latter covers a similar range of topics to this book.

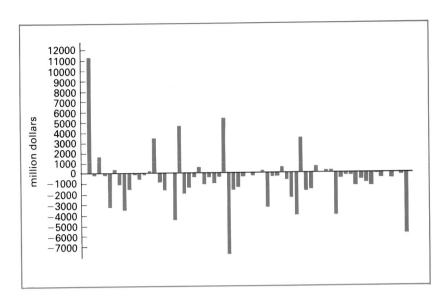

Fig. 1.6 Profitability of 65 randomly-selected US biotechnology companies. Data shown are for the second quarter (April–June) of 1990.

An up-to-date review of biotechnology but without the gene cloning emphasis is provided by I.J. Higgins, D.J. Best and J. Jones in their book, *Biotechnology, Principles and Applications* (1985, Blackwell Scientific Publications, Oxford). Coverage similar to this text can be found in J.L. Marx's book *A Revolution in Biotechnology* (Cambridge University Press, Cambridge).

More advanced coverage is provided by the three special issues of *Science*, the official journal of the American Association for the Advancement of Science: volume 196 number 4286 concentrated on the methodology of gene manipulation, volume 209 number 4463 on the research applications of the techniques and volume 219 number 4585 on commercial biotechnology. There are a number of journals devoted solely to biotechnology. The best of these are *Bio/technology* (Macmillan, London), *Trends in Biotechnology* (Elsevier, Amsterdam) and *Current Opinion in Biotechnology* (Current Science, London).

CITED REFERENCES

1 Hastings J.J.H. (1971) Development of the fermentation industries in Great Britain. *Developments in Applied Microbiology* **16**, 1–45.

Part II
Recombinant DNA Technology

2 The Basic Principles of Recombinant DNA Technology

Let us suppose that we wish to construct a bacterium that produces human insulin. Naïvely it might be thought that all that is required is to introduce the human insulin gene into its new host. In fact, a foreign gene is not maintained in its new environment if it is simply inserted into the bacterium on a DNA fragment for such fragments are not replicated (Fig. 2.1). The reason for this is that the enzyme DNA polymerase, which makes copies of the DNA, does not initiate the process at random but at selected sites known as *origins of replication*. Invariably, fragments of DNA do not possess an origin of replication. Using recombinant DNA technology it is possible to ensure the replication of the insulin gene in its new host by inserting the gene into a *cloning vector*. A cloning vector is simply a DNA molecule possessing an origin of replication and which can replicate in the host cell

of choice. Most cloning vectors used with microorganisms are extrachromosomal, autonomously replicating circles of DNA called *plasmids* (Fig. 2.1). Viruses are used occasionally as vehicles for gene insertion into microorganisms (see Chapter 4) but they are far more important for work with animal cells (see Chapter 9).

In order to insert foreign DNA into a plasmid, use is made of special enzymes known as *restriction endonucleases*. These enzymes cut large DNA molecules into shorter fragments by cleavage at specific nucleotide sequences called *recognition sites* (Fig. 2.2), i.e. restriction endonucleases are highly specific deoxyribonucleases (*DNases*). Some of these enzymes cut the two helices a few base pairs apart, generating two fragments with single-strand protrusions called *sticky ends* because their bases are complementary. Fragments of the foreign DNA are inserted into plasmid vectors cut open with the same enzyme or one which produces a matching end (Fig. 2.3). The resulting recom-

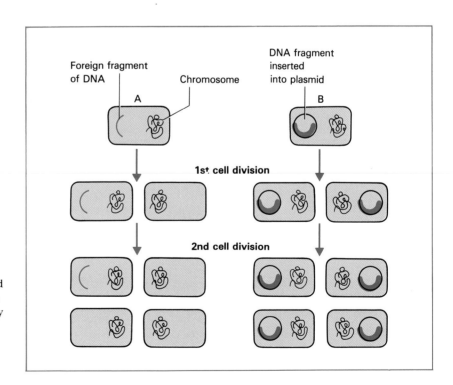

Fig. 2.1 The requirements for a cloning vector. (A) Fragments of DNA introduced into the bacterium by transformation do not undergo replication and are gradually diluted out of the population. (B) DNA fragments introduced into plasmids are inherited by both daughter progeny at cell division.

Enzyme	Recognition site		Cleavage products	
Eco RI	— GAATTC — — CTTAAG —		— G — CTTAA	AATTC — G —
Hae III	— GGCC — — CCGG —		— GG — CC	CC — GG —
Hind III	— AAGCTT — — TTCGAA —		— A — TTCGA	AGCTT — A —
Sma I	— CCCGGG — — GGGCCC —		— CCC — GGG	GGG — CCC —

Fig. 2.2 The recognition sites for some common restriction endonucleases. The arrows indicate the cleavage points. Note that the enzymes *Sma* I and *Hae* III generate blunt ends, while *Eco* RI and *Hind* III leave sticky ends.

Fig. 2.3 The construction of a chimeric (or recombinant) DNA molecule by joining together two DNA fragments produced by cleavage of different parental DNA molecules with the same restriction endonuclease.

binants or *chimeras* are introduced into the host cell by the process of transformation. In this process the DNA is mixed with a suspension of bacteria which have been prepared under specialized conditions, and the DNA enters the cell by a mechanism (*transformation*) which is still poorly understood.

PLASMIDS AS CLONING VECTORS

Hundreds of different cloning vectors, of which a representative selection is shown in Fig. 2.4, have been described and many of them have been constructed for special purposes. However, all of them have three features in common.

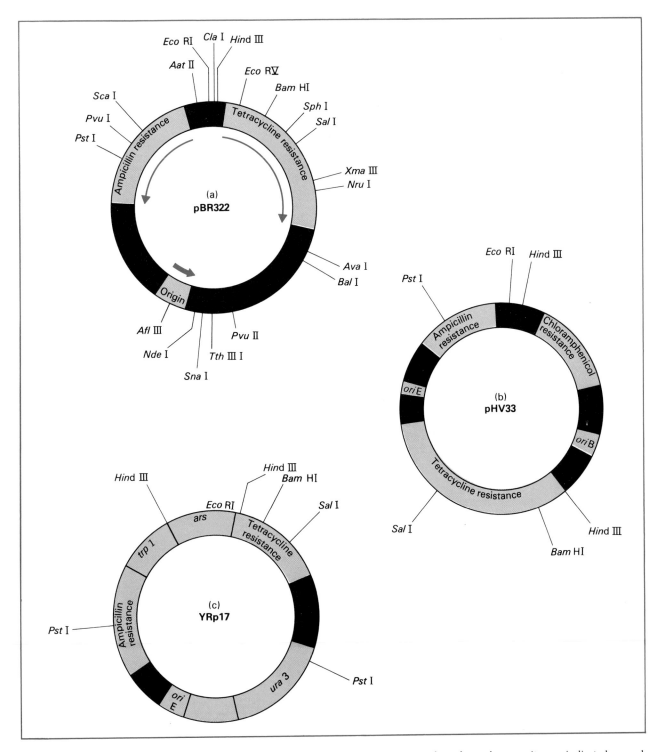

Fig. 2.4 Three plasmid vectors used for gene cloning (not drawn to scale). Restriction endonuclease cleavage sites are indicated around the perimeter of each vector. (a) The structure of pBR322, a vector used with *E. coli*. The thin arrows inside the circle show the direction of transcription of the two antibiotic resistance genes and the thick arrow the direction of DNA replication. (b) The structure of pHV33, a plasmid vector which can replicate in both *E. coli* and *B. subtilis*. *ori*E refers to the origin of replication necessary for plasmid maintenance in *E. coli* and *ori*B that for maintenance in *B. subtilis*. (c) A cloning vector which can replicate in both *E. coli* and yeast. The *ars* element is essential for replication in yeast. *trp* 1 and *ura* 3 are two yeast genes involved in the biosynthesis of tryptophan and uracil respectively. *ars* is a DNA sequence conferring autonomous replication of the vector in yeast cells.

1 The DNA of the cloning vector ideally has only a single target site for any particular restriction endonuclease. If it has more than one target site, bits of the plasmid can be lost or rearranged during the cloning process. The more restriction endonucleases for which the plasmid has unique sites then the better it will be as a vector.

2 All useful cloning vectors have one or more readily selectable genetic markers such as antibiotic resistance. The efficiency of the transformation process is very low and such markers are essential so that cells which have been transformed can be selected, e.g. on the basis of acquisition of ampicillin resistance.

3 As indicated above, one property of vectors is their possession of an origin of replication which ensures that they are propagated in the desired host cell. In some instances a plasmid is used as a cloning vector in two unrelated host cells, e.g. *Escherichia coli* and *Bacillus subtilis*. Such bifunctional vectors have more than one origin of replication (Fig. 2.4). Some cloning vectors used in yeast carry a centromere to facilitate segregation at cell division.

EXPRESSION OF CLONED GENES

Theoretically it is possible to clone in one organism any desired gene from another organism by the technique known as *shotgunning*. To do this the entire genome of the first organism is digested with a restriction endonuclease to produce a random mixture of fragments. These fragments are inserted into a plasmid vector and the recombinant plasmids transformed into the desired host cell. Since each recombinant plasmid will contain a different fragment of foreign DNA it is a major task (see p. 21) to select those transformed cells that carry the cloned gene of interest, in this case the human insulin gene. Even when the right cells can be identified, it is highly probable that they would not synthesize human insulin. That is, the information contained in the cloned gene is not used to create a functional protein or, in common parlance, the gene is not *expressed*.

There are a number of steps in gene expression:
1 transcription of DNA to mRNA;
2 translation of the mRNA into a polypeptide sequence;
3 in some instances, post-translational modification of the protein.
In many eukaryotes there is an additional step before translation, the removal of non-coding sequences from mRNA by a process known as splicing.

Transcription. Transcription of DNA into mRNA is mediated by the enzyme RNA polymerase. The process starts with RNA polymerase binding to recognition sites on the DNA which are called *promoters*. After binding, the RNA polymerase molecule travels along the DNA molecule until a termination signal is encountered. It follows that a gene which does not lie between a promoter and a termination signal will not be transcribed. Genes isolated in certain ways, such as by cDNA cloning or artifical synthesis (see p. 20), do not have their own promoter and they must be inserted into a vector close to a promoter site. An example is shown in Fig. 2.5. Even if a cloned gene carries its own promoter, this promoter may not function in the new host cell. In such circumstances the original promoter has to be replaced.

Translation. Translation of mRNA into protein is a complex process which involves interaction of the messenger with ribosomes. For translation to take place the mRNA must carry a *ribosome binding site* (rbs) in front of the gene to be translated. After binding, the

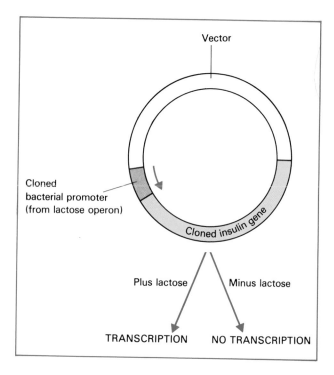

Fig. 2.5 Insertion of a cloned insulin gene into a vector carrying a bacterial promoter. The arrow within the circle indicates the direction of transcription. If the promoter is derived from the lactose operon, the transcription will be initiated only in the presence of lactose or a lactose analogue.

ribosome moves along the mRNA and initiates protein synthesis at the first AUG codon it encounters. It continues until it encounters a stop codon (UAA, UAG or UGA). If the cloned gene lacks a ribosome binding site it is necessary to use a vector in which the gene can be inserted downstream from both a promoter and an rbs (Fig. 2.6).

Splicing mRNA. Genes from bacteria and viruses have a very simple structure in that all the genetic information in the mRNA between the initiation and stop codons is translated into protein. Many genes of eukaryotic organisms, including the human insulin gene, have a more complex structure. They are a mixture of coding regions or *exons,* which contribute to the final protein sequence, and non-coding regions or *introns,* which are not translated into protein. In eukaryotes, genes containing introns are transcribed into mRNA in the usual manner but then the corresponding intron sequences are spliced out (Fig. 2.7). As bacteria cannot splice out introns they cannot be used directly to express many genes from mammals or other eukaryotes (Box 2·1).

One solution to the problem of introns is to clone the gene of interest in yeast (*Saccharomyces cerevisiae*), which can mediate splicing. Unfortunately cloning

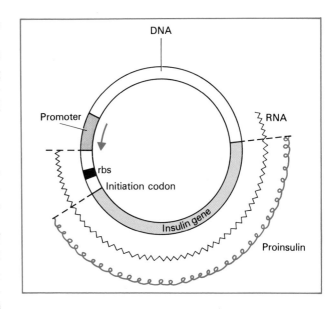

Fig. 2.6 The use of a vector carrying a promoter, an adjacent ribosome binding site (rbs) and an initiation codon to obtain synthesis of proinsulin from a synthetic gene. The arrow indicates the direction of transcription.

directly into yeast is much less efficient than cloning into a bacterium such as *E. coli* and this technique is not favoured. A better approach is to start by isolating

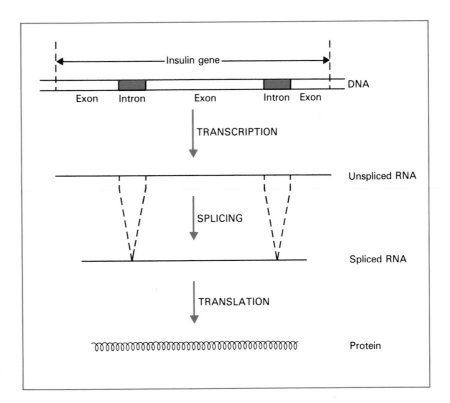

Fig. 2.7 Splicing of a messenger RNA molecule transcribed from a hypothetical insulin gene containing two introns.

Box 2·1 Ribozymes [1]

The term *ribozyme* is used to describe RNA molecules with catalytic activity. Enzymatic activities attributed to ribozymes include cleavage, annealing and ligation of nucleic acid substrates. Examples include self-splicing introns, the RNA component from RNase P and sequences derived from certain plant viruses.

In certain introns, the excision of the intron and subsequent ligation of the flanking exons to form a mature mRNA occurs by specific folding within the intron to produce an active unit that mediates splicing. RNase P is the enzyme responsible for cleavage of pre-tRNA molecules to mature tRNA. In all organisms studied so far, RNase P is made up of two subunits: an RNA and a protein but the RNA component can mediate cleavage in the absence of any protein. Autolytic cleavage in certain viroid and satellite RNAs associated with several plant viruses is an integral part of their replication cycle.

There has been some progress in adapting each of the natural ribozyme mechanisms described above in such a way that they are able to catalyse the cleavage of RNA molecules that are not their natural substrates. The eventual aim is to create ribozymes that can be targeted to specific RNAs *in vivo* as a means of controlling the expression of specific genes.

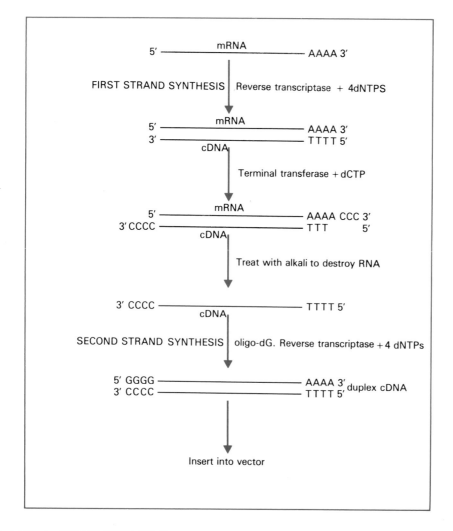

Fig. 2.8 Simplified method of using reverse transcriptase to convert mRNA into double-stranded DNA suitable for cloning.

mRNA from the original organism. In the case of insulin the mRNA would be obtained from human pancreatic cells, as these cells are rich in insulin-specific mRNA from which introns have already been spliced out. Using the enzyme *reverse transcriptase* it is possible to convert this mRNA into a DNA copy (Fig. 2.8). This copy DNA (*cDNA*), which carries the uninterrupted genetic information for insulin production, can then be cloned. It should be noted that such cDNA has *blunt ends* instead of sticky ends but it still can be inserted into plasmid vectors using blunt-end ligation (Fig. 2.9).

An alternative way of avoiding the problem of introns is to synthesize an artificial gene in the test-tube starting with deoxyribonucleotides (see Box 2·2, p. 20). In practice various oligonucleotides are synthesized and ligated together before insertion into an appropriate vector. This approach has been used to clone genes encoding proteins up to 500 amino

acids long but it demands that the entire amino acid sequence be known in advance.

Post-translational modifications. A number of proteins undergo post-translational modifications and insulin is one of these. Proteins that are destined to be transported out of the cell are synthesized with an extra 15—30 amino acids at the amino-terminus (N-terminus). These extra amino acids are referred to as a *signal sequence* and a common feature of these sequences is that they have a central core of hydrophobic amino acids flanked by polar or hydrophilic residues. During passage through the membrane the signal sequence is cleaved off (Fig. 2.10). If our insulin gene were cloned by the cDNA method, the signal sequence would be present and, in *E. coli* at least, the insulin would be transported through the cytoplasmic membrane (*exported*). Using the synthetic gene approach a signal sequence would be present on the protein only if the

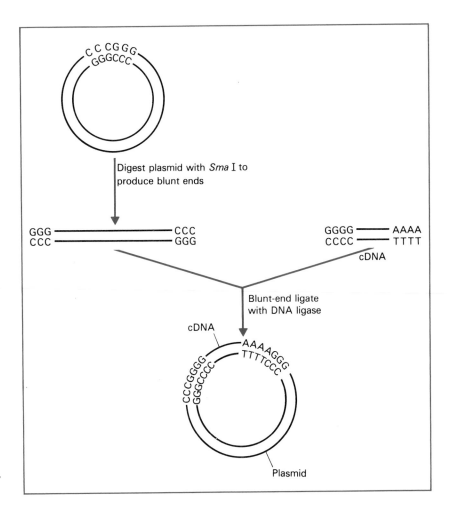

Fig. 2.9 The use of blunt-end ligation to clone a cDNA fragment. The DNA sequence shown on the plasmid is the recognition site for restriction endonuclease *Sma* I (see Fig. 2.2). The cDNA shown is the one generated in Fig. 2.8.

Box 2·2 The chemical synthesis of oligonucleotides and genes [4]

The ability to synthesize genes chemically is fundamental to the success of modern biotechnology. Not only does it enable complete genes to be synthesized to facilitate cloning, as described in this chapter, but it enables minor or major changes to be made in the gene product. This latter aspect, protein engineering, is discussed in Chapter 4. In addition, short oligonucleotides, which are easy to synthesize, can be used diagnostically in a variety of ways as described in Chapter 3.

The basic method of gene synthesis is the repetitive formation of an ester linkage between an activated phosphoric acid function of one nucleotide and the hydroxyl group of another nucleoside or nucleotide, thus forming the characteristic phosphodiester bridge. The major problem is that deoxyribonucleotides are very reactive molecules, having a primary and secondary hydroxyl group, a primary amino group and a phosphate group. Consequently, blocking and deblocking procedures are required and the chemistry involved must not result in scission or alteration of the phosphodiester backbone, the furanose rings, the sugar—purine/pyrimidine bond or the bases themselves — a tall order indeed!

Currently, the most popular method for the synthesis of short nucleotides (called *oligonucleotides*) makes use of the extreme reactivity of phosphite reagents. The basic principles of the method are shown in Diagram 1. Although the method will work in solution for the condensation of three or four nucleotides, construction of large oligonucleotides requires the 3' end of the desired oligonucleotide to be coupled to an insoluble support. Fortunately, immobilization simplifies manipulative procedures and cuts out time-consuming purification steps following each cycle of condensation. Appropriately blocked mononucleotides are added sequentially and reagents, starting materials and by-products removed by filtration. At the conclusion of the synthesis the deoxyoligonucleotide is chemically freed of blocking groups, hydrolysed from the support and purified by electrophoresis or high-performance liquid chromatography (HPLC). By immobilizing the polymer support carrying the initiating deoxynucleotide in a column, the filtration steps can be replaced by a simple washing procedure and this lends itself to a fully automatic synthesis. Automatic gene synthesizers consist simply of reagent reservoirs whose contents are added to or removed from the immobilized protected oligonucleotide via valves controlled by a microprocessor.

X = Solid support anchorage position
Y = Dimethoxytrityl
Z = Methyl

Diagram 1

Fig. 2.10 The conversion of preproinsulin to insulin by sequential removal of the signal peptide and the C fragment.

nucleotide sequence corresponding to the protein's signal sequence had been incorporated at the time of gene construction. Sometimes it is desirable for the bacterium to export the protein, in which case a signal sequence is incorporated; in other cases it may be desirable that proteins are retained within the cell.

In *E. coli* cells the presence of a signal sequence does not usually result in export of a protein into the growth medium. Instead it is directed to the periplasmic space, that is the space between the cytoplasmic membrane and the complex membrane-like cell wall. Unfortunately, many recombinant proteins are rapidly and extensively degraded in the periplasmic space due to the presence there of numerous proteolytic enzymes (*proteases*). In Gram-positive species such as *Bacillus*, and in eukaryotic microorganisms, signal sequences direct the export of proteins into the growth medium.

The export process is even more complex in that protein synthesis and export occur simultaneously on the inner surface of the cell membrane. If the rate of synthesis of an exported protein is very high, it is possible to envisage a situation in which the membrane becomes 'jammed' with protein in the various stages of synthesis and export. This problem can be minimized by using as a host cell for the recombinant plasmid one which has a very high surface area to volume ratio. Suitable organisms are actinomycetes and filamentous fungi.

A small number of proteins, and again insulin is a good example, are synthesized as *proproteins*. This means that there is an additional amino acid sequence which dictates the final three-dimensional structure but which is deleted before the protein becomes fully functional (Fig. 2.10). In the case of proinsulin, proteolytic attack cleaves out a stretch of 35 amino acids in the middle of the molecule to generate insulin. The

peptide that is removed is known as the C fragment. The other chains, A and B, remain cross-linked and thus locked in a stable tertiary structure by the disulphide bridges formed when the molecule originally folded as proinsulin. Bacteria have no mechanism for specifically cutting out the folding sequences from proproteins. If native insulin is to be made using recombinant bacteria, it is necessary to synthesize the A and B chains separately, purify them and join them together *in vitro*.

Another post-translational modification occurring naturally in the construction of many proteins is the addition of oligosaccharides to certain amino acid residues. This process of *glycosylation* occurs, for example, in the synthesis of the antiviral interferon-β and -γ. Bacteria cannot glycosylate the products of cloned mammalian genes. Yeast cells can glycosylate proteins but the pattern of glycosylation differs from that mediated by animal cells. The problems associated with no or incorrect glycosylation are discussed in more detail later (p. 94). Where the correct glycosylation is necessary the only option is to clone the corresponding gene in a mammalian cell (see Chapter 9).

SELECTION OF RECOMBINANTS

The task of isolating a desired recombinant from a population of transformed bacteria depends very much on the cloning strategy that has been adopted; for instance, if a synthetic gene has been cloned, no selection is necessary because every transformed cell will contain the correct sequence. When a cDNA derived from a purified or abundant mRNA is to be cloned, the task is relatively simple: only a small number of clones needs to be screened. Isolating a particular single-copy gene sequence from a complete mammalian genomic library requires techniques in which hundreds of thousands of recombinants can be screened. The advantages and disadvantages of the different cloning strategies are shown in Table 2.1.

A number of different methods have been devised to facilitate screening of recombinants. These include genetic methods, immunochemical methods and methods based on nucleic acid hybridization. The simplest example of a genetic method is the complementation of nutritional defects; for example, suppose a bacterial strain is available which has a mutation in a gene encoding an enzyme involved in the biosynthesis of the amino acid histidine. Such a mutant strain will only grow in medium supplemented with histidine. By cloning DNA from a normal strain, i.e. one that can synthesize its own histidine, in the mutant strain and selecting those transformants which grow in the absence of histidine it is possible to isolate the gene of interest. Even yeast and mammalian genes can be selected in this way.

In a procedure analogous to that described above, a clone carrying the mouse dihydrofolate reductase (DHFR) gene was selected from a population of recombinant plasmids containing cDNA derived from an unfractionated mouse cell mRNA preparation. In this instance the basis of selection was that mouse DHFR is more resistant to the inhibitor trimethoprim than *E. coli* DHFR. When trimethoprim was added to the medium on which transformed *E. coli* were grown, only those cells carrying the mouse DHFR gene survived.

Immunochemical methods are widely used to select particular recombinant clones but demand the availability of a specific antibody against the desired protein product. When such an antibody is available the method works well. Transformed cells are grown on agar in a conventional Petri dish and a duplicate set of colonies prepared in a second Petri dish. The cells in the duplicate set of colonies are lysed by exposure to chloroform vapour and the released proteins blotted on to an adsorbent matrix. In effect, this produces

Cloning method	Advantages	Disadvantages
Shotgun cloning	1 Easy to do	1 Many different clones isolated so need good selection method to isolate desired clone 2 Genes with introns will not be correctly expressed in *E. coli*, the usual shotgun cloning host 3 Expression dependent on recognition of foreign promoters by *E. coli* 4 Codon choice may not be optimal for *E. coli*
cDNA cloning	1 If desired mRNA is in abundance in the mRNA preparation used, desired clone should be easily detected 2 Introns not a problem as mRNA will contain spliced molecules	1 Desired mRNA not always abundant so selection may be necessary 2 Technically more difficult than shotgun cloning 3 Cloned gene needs to be placed downstream from a promoter, since normal promoter eliminated by starting with mRNA 4 Codon choice may not be optimal for *E. coli*
Gene synthesis	1 No selection needed after cloning 2 Sequence of promoters, ribosome binding sites, etc., can be optimized 3 Codon choice can be optimized for preferred host cell	1 Need to know protein sequence before a synthetic gene can be designed

Table 2.1 Comparison of the different cloning strategies.

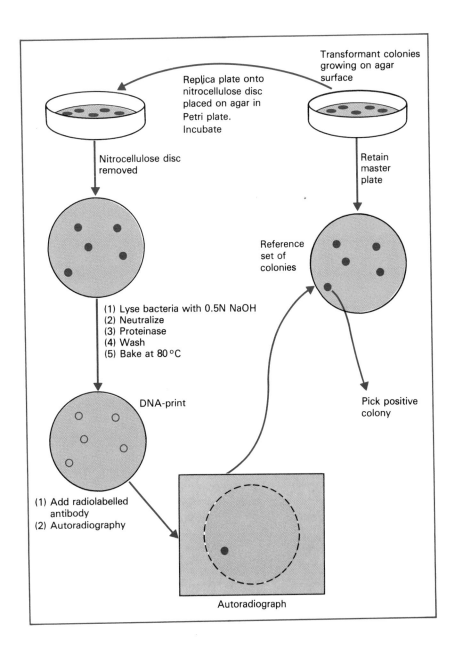

Replica plate onto
nitrocellulose disc
placed on agar in
Petri plate.
Incubate

Transformant colonies
growing on agar
surface

Nitrocellulose disc
removed

Retain
master
plate

(1) Lyse bacteria with 0.5N NaOH
(2) Neutralize
(3) Proteinase
(4) Wash
(5) Bake at 80 °C

Reference
set of
colonies

Pick positive
colony

DNA-print

(1) Add radiolabelled
    antibody
(2) Autoradiography

Autoradiograph

Fig. 2.11 The basic method of immunological screening of recombinants.

a map of the original colonies in the form of their protein products. The matrix is then exposed to antibody which has been radioactively labelled *in vitro*. Positively-reacting lysates are detected by washing surplus radiolabelled material off the matrix and making an autoradiographic image (Fig. 2.11). Many different variations of this technique have been adopted.

Detection methods based on nucleic acid hybridization are also widely used but these techniques have other applications as well and are described in more detail in Chapter 3.

SPECIFIC DNA AMPLIFICATION: THE POLYMERASE CHAIN REACTION (PCR) [2, 3]

The PCR technique, which was first reported in 1985, has revolutionized gene manipulation procedures because it enables specific DNA sequences to be selectively amplified a thousand- or even a million-fold. PCR amplification involves two oligonucleotide primers that flank the DNA segment to be amplified. These primers hybridize to opposite strands of the target sequence (Fig. 2.12) and are oriented so that DNA synthesis by the polymerase proceeds across the regions between the primers, effectively doubling the amount of that DNA segment. Since the extension products are also complementary to primers, the cycle can be repeated after a denaturation step. By repeated cycles of denaturation, priming and extension there is a rapid exponential accumulation of the specific target fragment. Thus after only 10 cycles the target sequence will have been amplified 1000-fold. To simplify the technique the DNA polymerase used is derived from the thermophile *Thermus aquaticus* so that it does not get inactivated during the denaturation step.

Although PCR amplification has many uses only two will be described here. Others will be discussed in subsequent chapters. In terms of cloning genes, the PCR technique can eliminate the need for sensitive screening systems like those described in the previous section. Rather, the gene to be cloned is first amplified *before* it is inserted into the cloning vector. Another use is to recombine DNA molecules at precise junctions irrespective of nucleotide sequences at the recombination site and without the use of restriction endonucleases or ligase. The principle of the method is shown in Fig. 2.13.

MAXIMIZING GENE EXPRESSION

If the objective of cloning a mammalian gene in a microorganism is to facilitate commercial production of the corresponding gene product, it follows that it

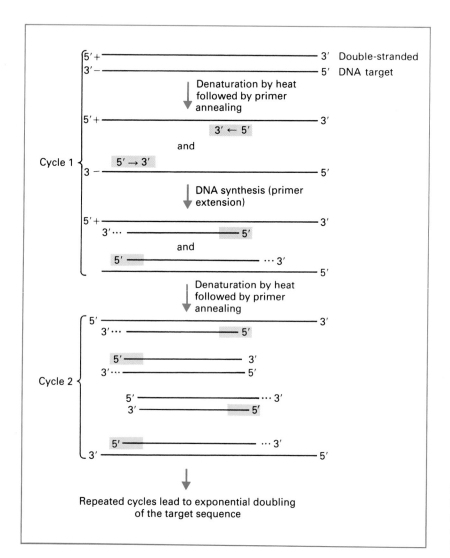

Fig. 2.12 The polymerase chain reaction. Note that although the amplified DNA is not a full copy of the starting DNA it can still bind fresh primer and so participate in further amplification steps.

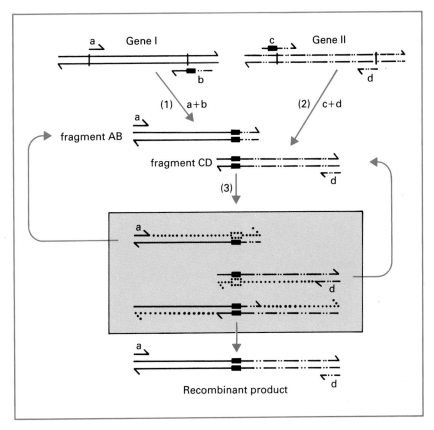

Fig. 2.13 Mechanism of gene splicing by overlap extension (SOE). Segments from two unrelated genes are to be spliced. Gene I is shown as solid lines and Gene II as dashes and dots. The 5′−3′ direction of each strand is shown by half arrow-heads. Oligos are labelled with single lower-case letters, and PCR products are labelled using the two corresponding upper-case letters; for example, reaction (1) is a PCR reaction using primers 'a' and 'b' on gene I to generate the AB product. Product CD is generated in a separate reaction. Oligos 'b' and 'c' match their template genes in their 3′ portions, but their 5′ portions are designed so that the two oligos are complementary. The two PCR products are shown with the homologous segments aligned. In reaction (3), the segments are mixed along with excess primers 'a' and 'd', denatured, reannealed, and primer-extended by DNA polymerase (PCR conditions). The intermediates in this reaction are shown in the hatched box. The end of one strand from each product is capable of hybridizing with the complementary end from the other product. The strands having this overlap at their 3′ ends can act as primers for one another and be extended by the polymerase to form the full-length recombinant product; DNA segments synthesized by the polymerase are shown as dotted lines. The recombinant product is PCR-amplified in the presence of 'a' and 'd'.

is essential to maximize gene expression. Important factors are:

1 the number of copies of the plasmid vector per unit cell (*copy number*);

2 the strength of the promoter;

3 the sequence of the rbs and flanking DNA;

4 codon choice in the cloned gene;

5 genetic stability of the recombinant;

6 proteolysis.

The limiting factor in expression is the initiation of protein synthesis. Increasing the number of plasmids per cell increases the number of mRNA molecules transcribed from the cloned gene and this results in increased protein synthesis. Similarly, the stronger the promoter, the more mRNA molecules are synthesized. The nucleotide base sequence of the rbs and the length and sequence of the DNA between the rbs and the initiating AUG codon are so important that a single base change, addition or deletion can affect the level of translation up to 1000-fold.

Another important factor is related to the redundancy of the genetic code. There are several trinucleotide codons for most amino acids and different organisms favour different codons in their genes. If genes inserted into cells of another species utilize codons rare in the host cell, the host's biosynthetic machinery may be starved of charged tRNAs. This could result in premature protein chain termination or a high error frequency in the amino acid sequence of the protein.

In any culture containing cells engaged in excess synthesis, be it of a protein or a low-molecular-weight molecule (e.g. an antibiotic), there is a strong selective pressure in favour of non-producing cells. Cells which are not encumbered by extra production can invest their energy in cell division and hence rapidly out-number producers. Non-producing cells frequently arise in transformed cultures as a result of spontaneous changes in gene sequence, e.g. deletions and gene rearrangements. To minimize selection against cells which can achieve high product levels it is wise to minimize recombinant gene expression until the final production vessel is reached. This can be achieved by using controllable promoters and plasmid vectors with controllable copy number. An example of a controllable promoter is that derived from the *E. coli* lactose operon: no transcription occurs unless lactose is added (Fig. 2.5). *Runaway plasmids* are examples of vectors with controllable copy number. At low temperatures, e.g. below 30 °C, the copy number may be as low as 10 plasmids per cell but when the temperature is raised to 37 °C the copy number increases to several hundred, or even several thousand.

The enzymatic breakdown of protein (*proteolysis*) does not affect transcription and translation but by degrading the desired product it influences the *apparent* rate of gene expression. Although proteolysis can be reduced, it is difficult to eliminate completely. One approach which is used widely is to 'protect' the desired protein by fusing it to a normal cellular protein from which it must subsequently be released. This approach is discussed in more detail in Chapter 4.

Further reading

GENERAL REFERENCES

Old R.W. & Primrose S.B. (1990) *Principles of Gene Manipulation*, 4th edn. Blackwell Scientific Publications, Oxford.

CITED REFERENCES

1 Perriman R.J. & Gerlach W.L. (1990) Manipulating gene expression by ribozyme technology. *Current Opinion in Biotechnology* 1, 86–91.
2 Van Brunt J. (1990) Amplifying genes: PCR and its alternatives. *Bio/technology* 8, 291–3.
3 Horton R.M., Hunt H.D., Ho S.N., Pullen J.K & Pease L.R. (1989) Engineering hybrid genes without the use of restriction enzymes: gene splicing by overlap extensions. *Gene* 77, 61–68.
4 Davies L., Johnson I.D. & Davies J.A. (1991) Designer genes — design and synthesis illustrated for the gene coding for human macrophage colony stimulating factor. In *Genetic Manipulation: Techniques and Applications* (Ed. by J.M. Grange, A. Fox & N.L. Morgan), pp. 351–359. Blackwell Scientific Publications, Oxford.

3 Recombinant DNA Technology and the New Diagnostics

The production of reagents for the diagnosis of biological disorders and infectious disease is a major industry and on a global scale revenue from sales is estimated at many billions of dollars. Traditionally diagnostic reagents have taken two forms:

1 biochemical reagents for assaying specific enzymes;
2 antibodies for detecting specific proteins by such techniques as immunofluorescence, radioimmunoassay (RIA) and enzyme-linked immunosorbent assay (ELISA).

Although the former are used principally in clinical biochemistry, the latter have applications in almost all industries, e.g. health care, food and agriculture. Recently a number of new types of diagnostic reagent have been developed. The first of these relies on the use of DNA or RNA probes to detect particular gene sequences. As will be discussed later, applications of the technique include the identification of hereditary defects in embryos and the identification of particular plant diseases. Although based on an old technique, nucleic acid hybridization, development of these reagents awaited the availability of milligram quantities of specific genes. The ready provision of such large amounts of gene sequences is easy if the sequences have been cloned in a vector, since a 1l bacterial culture can yield up to half a milligram of plasmid DNA.

For hybridization of a nucleic acid probe to a test gene to be detected or quantified some method of labelling the probe is required. In the past, probes with a radioactive label were prepared from bacteria grown in the presence of radioisotopes. More recently, methods of labelling probes *in vitro* have been devised.

LABELLING NUCLEIC ACIDS *IN VITRO* [1]

One method of labelling DNA gene probes *in vitro* is by means of a technique known as *nick translation* (Fig. 3.1). In any preparation of DNA, particularly one that is more than a few days old, random single-strand breaks ('nicks') occur. Starting at the free 3'-hydroxyl group exposed in these nicks, the enzyme DNA polymerase I (see Box 3·1) will incorporate free nucleotides. Concomitantly the enzyme digests away single-stranded DNA from the exposed 5' end. If one or more of the deoxynucleoside triphosphates is labelled in the α position with phosphorus-32, the reaction progressively incorporates the radioactive label into a duplex that is unchanged except for movement ('translation') of the nick along the molecule. DNA probes of high specific activity (10^8 disintegrations per min (dpm)/μg DNA) can be obtained readily by this method.

An alternative method of labelling DNA is to use the *random primer method*. The basic principle of the method is shown in Fig. 3.2. A mixture of random-sequence hexanucleotides is added to DNA. Some of those hexanucleotides will hybridize to the DNA and act as initiation sites for DNA polymerase. When the four deoxynucleotide triphosphates are added, only one of which need be radioactively labelled, probes 200–400 nucleotides in length are produced. These probes can be released by denaturation and used for hybridization without purification. By this method, probes can be produced whose specific activity is ten times higher than those prepared by nick translation.

Single-stranded RNA can also be used as a gene probe. Indeed, it is probably the most sensitive probe for any diagnostic procedure based on hybridization. This is because the rate and stability of hybridization is far greater in an RNA–DNA hybrid than in a DNA–DNA hybrid. The best system for labelling RNA is the SP6 system (Fig. 3.3) since it readily yields specific activities in excess of 5×10^9 dpm/μg RNA. The gene from which the probe is to be prepared is cloned in a plasmid vector such that it is under the control of the promoter of the phage SP6 RNA polymerase gene. After purification the plasmid is linearized with a suitable restriction enzyme and then incubated with SP6 RNA polymerase and the four ribonucleoside triphosphates. If one of the triphosphates is radioactively labelled, the incubation mix will contain labelled RNA, which can be added directly to the hybridization mixture.

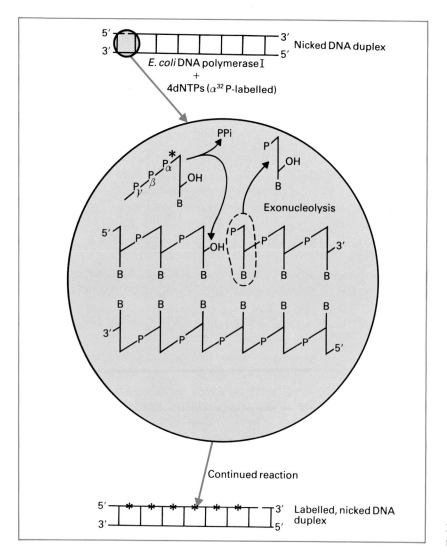

Fig. 3.1 Synthesis of labelled DNA by nick translation.

Example: Detection of viroids in plants [2]. Viral infection of crop plants can have serious agronomic consequences, not the least of which is decreased vigour and reduced crop yield. Preventive measures include the cultivation of crops from virus-free stocks and the prevention of these pathogens from entering and spreading through crops. In these procedures suitable diagnostic tests for the rapid and reliable detection of viruses are of paramount importance. Enzyme-linked immunosorbent assays (ELISA) have been developed for the detection of a number of important plant viruses and many of these are in widespread use. Viroids cause a number of important plant diseases (see Box 3·2) but because they lack the antigenic protein coat characteristic of viruses they cannot be detected by immunological procedures. However, they can be detected by probing plant extracts with a radioactively labelled plasmid carrying a copy of the viroid genome. The method is shown in Fig. 3.4.

Test plants are homogenized in a buffer and the resulting extract clarified by centrifugation. The clarified extract is spotted on to a nitrocellulose membrane and dried in a vacuum, which causes the viroid RNA to bind irreversibly. Binding occurs by means of the sugar–phosphate backbone of the RNA and leaves the bases available for hydrogen bonding to complementary bases. An immediate problem arises in that added probe DNA will also bind to the filters. To avoid this problem any nucleic acid binding sites remaining on the nitrocellulose after the test RNA has bound are blocked with bovine serum albumin. When the filters have been prepared the radiolabelled probe

Box 3·1 DNA polymerase

A number of DNA polymerases from eukaryotic and prokaryotic sources have been well characterized. Most of those from eukaryotes possess only a DNA synthesizing activity, whereas the corresponding bacterial enzymes also possess exonuclease activity. There are three DNA polymerases in *E. coli* of which DNA polymerase I is the most versatile and useful. DNA polymerase I is a single polypeptide of molecular weight 109 000 and possesses three activities: a DNA synthesizing activity, a 3′ → 5′ exonuclease activity and a 5′ → 3′ exonuclease activity. The enzyme can be cleaved into two fragments by protease treatment. The larger fragment, known as the *Klenow fragment*, has a molecular weight of 76 000 and contains the polymerase and 3′ → 5′ exonuclease activities. The small fragment (mol-ecular weight 36 000) has the 5′ → 3′ exonuclease activity.

DNA polymerase I has the unique ability to initiate replication *in vitro* at a nick in DNA and to extend the 3′-OH end. As the new segment of DNA is synthesized, it displaces the existing homologous strand in the duplex. The displaced strand, in turn, is digested by the 5′ → 3′ exonuclease activity. This process of *nick translation* (Diagram 1) is used to radiolabel DNA *in vitro* (see main text).

All of the known DNA polymerases can extend a DNA chain only from a free 3′-OH end. When presented with a single-stranded template none can initiate synthesis unless a start primer is added to present the requisite 3′-OH group (Diagram 2). These primers may be either RNA or DNA.

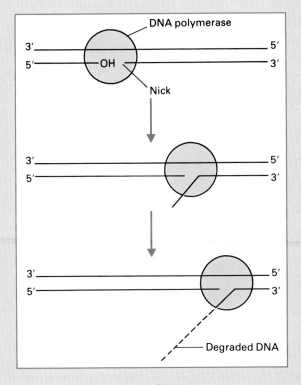

Diagram 1 Nick translation of DNA.

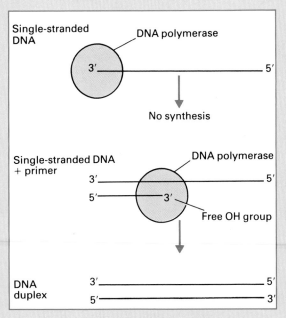

Diagram 2 The need for a primer for DNA synthesis.

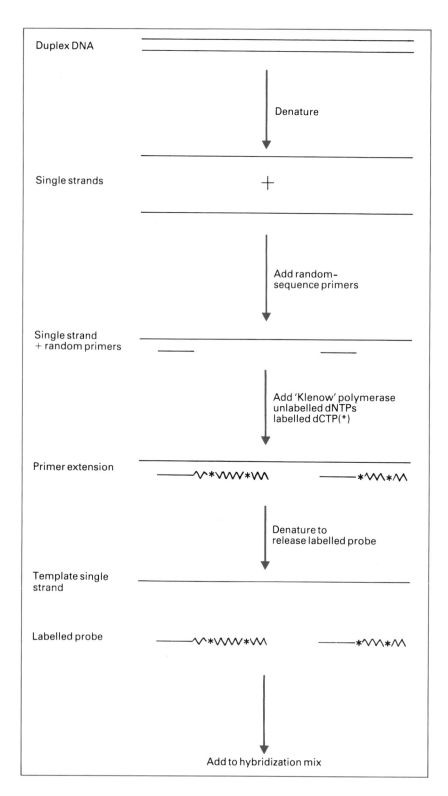

Duplex DNA

Denature

Single strands

+

Add random-
sequence primers

Single strand
+ random primers

Add 'Klenow' polymerase
unlabelled dNTPs
labelled dCTP(*)

Primer extension

Denature to
release labelled probe

Template single
strand

Labelled probe

Add to hybridization mix

Fig. 3.2 The random primer method for preparing labelled DNA.

is added and the mixture incubated under conditions appropriate for hybridization. The filter is washed to remove excess probe which has not hybridized and any positive samples are identified by autoradiography.

Probes of the kind described above can also be used to facilitate plant breeding; for example, one goal of potato breeding programmes is the development of cultivars which are resistant to virus diseases transmitted by aphids. Identification of suitable plants is effected by exposing them to virus-bearing aphids and then testing the cell sap with DNA probes for potato viruses X and Y and potato leafroll virus. A different application is in wheat breeding. Many European wheats have a chromosome IB in which the short arm is derived from rye. Some breeders are interested in selecting plants in which this chromosome is entirely from wheat. By using a rye DNA sequence as a probe, plants with IBs from rye can be detected simply by squashing a root tip on to a nitrocellulose filter and then applying a suitable probe.

USE OF GENE PROBES VERSUS IMMUNOLOGICAL DETECTION [1]

The use of gene probes is not restricted to the detection of viroids; with a slight modification they can also be used to detect bacteria. In this case the test sample is applied to a nitrocellulose filter as before. If the filter is dropped into sodium hydroxide solution, the bacteria will lyse, the released DNA will denature and the separated strands will bind loosely to the nitrocellulose. After washing to remove the alkali, the filter is dried in a vacuum oven as before. Probing is then carried out as described in the previous section.

Although a large number of different microorganisms can now be detected using gene probes, the method has three disadvantages when compared with immunological methods. The first of these is the use of radioactive labels, although non-radioactive methods of labelling nucleic acids are being developed and these are described in the next section. The second disadvantage is the overall test time, which is usually 24–72 h. With the use of non-radioactive labelling methods this may be reduced to 6–12 h but even this is slower than a serological method. The third disadvantage relates to the use of probes to detect microbes in foodstuffs which have been heat-treated: if a positive result is obtained there is no way of knowing if the organisms are alive or dead.

On the positive side, probe technology has a number of advantages over immunological detection methods. First, the method can be applied without modification to samples of blood, faeces, pus or body exudates which are too dirty to permit antibody–antigen interaction. The alkali denaturation step effectively cleans the sample. Second, probe technology can detect pathogenicity determinants which would not be revealed immunologically. For example, many of the cases of diarrhoea and vomiting which occur amongst persons travelling abroad, the so-called 'traveller's tummy', are caused by strains of *Escherichia coli* which secrete an enterotoxin. Whereas the genes for enterotoxin are plasmid-borne and are detected easily with appropriate probes, there are no differences between toxigenic and non-toxigenic *E. coli* which can be detected with antisera. Third, if sufficient of the infectious agent is present then no cultivation of the microbe is necessary. This eliminates hazards. The polymerase chain reaction is particularly powerful in this respect. For example with PCR technology it is possible to detect HIV-1 DNA in peripheral blood mononuclear cells in

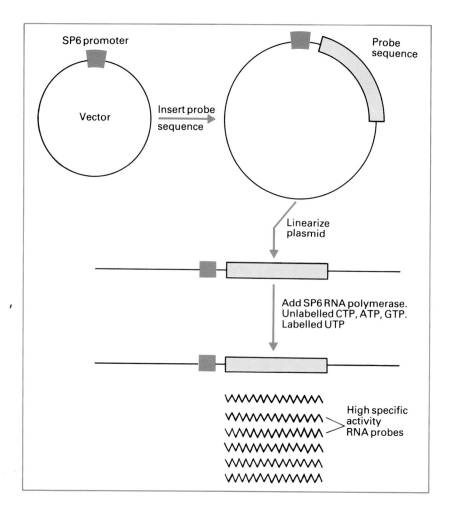

Fig. 3.3 The SP6 system for preparing labelled RNA.

less than 1 day whereas virus isolation takes 3–4 weeks. Finally, hybridization technology can be used even when the target organism cannot be cultured. This is often the case with viral infections where the current diagnostic method is to look for rising titres of antibody in the blood. The method will also work with latent infections when no antibody may be present. The probes currently approved for use by the US Food and Drug Administration are shown in Table 3.1.

An important diagnostic tool is *in situ* hybridization (ISH). In most instances ISH is carried out on formalin-fixed, paraffin-embedded tissues. This technique has proved particularly useful for the detection of viral pathogens. It allows one to examine the tissue first by traditional staining methods and then to correlate the cytology seen with the presence of infectious agents. The technique should prove particularly useful in diagnosing those cells which have become cancerous as the result of gene changes.

Table 3.1 Probes approval for the diagnosis of microorganisms.

Bacteria
Legionella
Mycobacterium tuberculosis complex
Mycobacterium spp.
Mycobacterium avium-supracellulare
Mycoplasma pneumoniae
Chlamydia
Campylobacter spp.

Viruses
Herpes simplex types 1 and 2

NON-RADIOACTIVE LABELS FOR PROBES [1, 3]

Although probes labelled with radioactive phosphorus are convenient for the research worker, they are not particularly suitable for a diagnostic laboratory where quality control and long shelf-life are important for

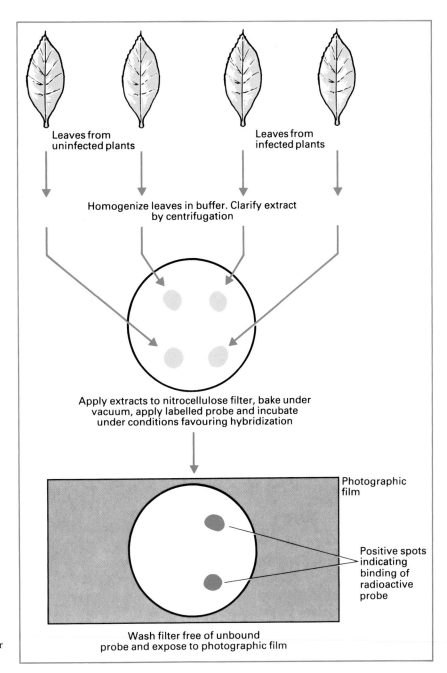

Leaves from
uninfected plants

Leaves from
infected plants

Homogenize leaves in buffer. Clarify extract
by centrifugation

Apply extracts to nitrocellulose filter, bake under
vacuum, apply labelled probe and incubate
under conditions favouring hybridization

Photographic
film

Positive spots
indicating
binding of
radioactive
probe

Wash filter free of unbound
probe and expose to photographic film

Fig. 3.4 Use of a gene probe to detect plants infected with a viroid. See text for details.

standardization of tests. By using radioactively labelled [^{35}S]- thionucleotides the shelf-life of the probes can be extended.

An alternative solution is to use non-radioactive labels and two basic methods have been devised. In *indirect* labelling the probe nucleic acid is modified by the incorporation of analogues (e.g. biothin- or digoxigenin-labelled nucleotides) during synthesis. The presence of the modified nucleotides is detected by the binding of a signal generating system, such as alkaline phosphatase, via an avidin or antibody coupling as shown in Fig. 3.5. In *direct* labelling the enzyme reporter molecule is directly cross-linked to the nucleic acid. For example horseradish peroxidase or alkaline phosphatase can be cross-linked to polyethyleneimine with *p*-benzoquinone and the resulting conjugates

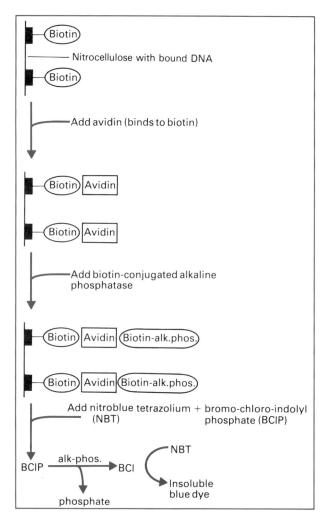

Fig. 3.5 The detection of biotin-labelled DNA using avidin and alkaline phosphatase. The insoluble blue dye produced by the reduction of nitroblue tetrazolium 'stains' the nitrocellulose in the vicinity of biotin-labelled DNA.

covalently linked to DNA using glutaraldehyde (Fig. 3.6). Because of the simple chemistry involved, direct labelling is much quicker than indirect labelling.

SOUTHERN BLOTTING [4, 5, 6, 7]

The diagnostic potential of DNA−DNA hybridization can be increased greatly by means of *Southern blotting*. This technique is named after Ed Southern, who developed it in 1975. The basic principle of the method is shown in Fig. 3.7. A sample of DNA is digested with a restriction endonuclease and the fragments which are produced are separated on a size basis by electrophoresis in an agarose gel. The DNA

fragments are then denatured into single strands by soaking the gel in alkali. Subsequently, the gel is placed on a wick of filter paper that is immersed in a trough of buffer. A sheet of nitrocellulose membrane is placed on top of the gel and a large stack of absorbent paper (usually paper towels) on top of this. The buffer solution is drawn up by the absorbent paper, passes through the gel and nitrocellulose and carries the DNA out of the gel and on to the membrane. As before, the DNA is fixed in place by drying the filter in a vacuum. The net result is a replica on the nitrocellulose of the DNA fragment pattern from the agarose gel.

Example: Prenatal detection of inherited defects. Amniocentesis is a technique whereby a syringe is carefully inserted through the abdomen of a pregnant woman and used to withdraw a sample of the amniotic fluid which surrounds the fetus. This amniotic fluid contains some cells (amniocytes) derived from the developing child and, if desired, these can be cultured *in vitro* to increase their number. Alternatively, the DNA in the amniocytes can be amplified using the PCR. Since, with the exception of the antibody-producing B lymphocytes, the DNA in every somatic cell is identical, these amniotic cells are an accurate reflection of the genetic composition of the fetus. Using cloned gene probes it is possible to determine whether the fetus is carrying a particular inherited metabolic disorder and subsequently arrangements can be made for an abortion, if so desired.

The easiest form of gene defect to detect is where the loss of gene function is caused by a deletion of part or all of a gene sequence resulting in a reduction in size of a particular DNA restriction fragment in the Southern blot. This has occurred in a number of thalassaemias (globin disorders) and certain types of osteogenesis imperfecta (a collagen disorder). Many other inherited diseases are not due to deletions but to single-base changes in a gene that result in a protein with an altered amino acid sequence. Sometimes this change of DNA sequence can be detected because it creates or destroys a site for a particular restriction enzyme. A good example of this is the sickle-cell mutation where the change in DNA sequence destroys a site for the restriction enzyme *Mst* II leading to a different pattern of β-globin fragments on Southern blotting (Fig. 3.8).

Since this method of detecting *restriction fragment length polymorphisms* (RFLP) is limited by the sequence specificities of restriction enzymes, a more sophisti-

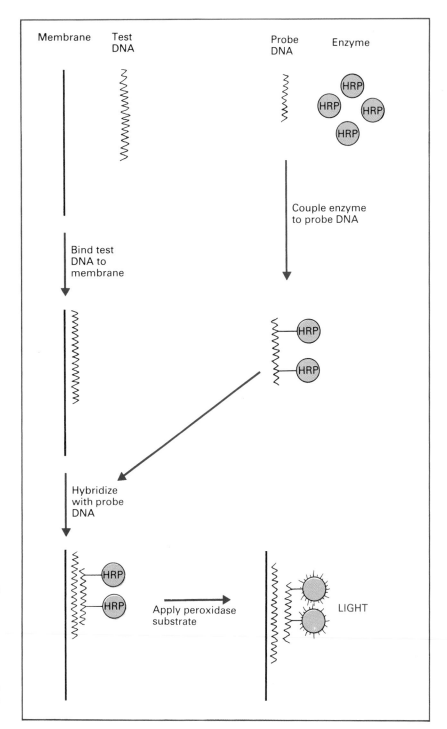

Fig. 3.6 The use of direct-labelled probes to identify gene sequences. The enzyme horseradish peroxidase is coupled to single-stranded probe DNA using polyethyleneimine, *p*-benzoquinone and gluteraldehyde. After hybridization the addition of a suitable substrate results in light formation which is detected using a suitable photographic film. See reference [3] for more details.

cated method has been developed. In this approach two oligonucleotides, e.g. of 19 bases long, are chemically synthesized. One is perfectly complementary to the normal DNA sequence around the site of the mutation and the other is complementary to the mutant sequence. These two oligonucleotides, which differ in only one base out of 19, can be radiolabelled *in vitro* and used as probes in a Southern blot analysis of normal, carrier and patient DNA samples. By washing the blots at different temperatures after hybridization,

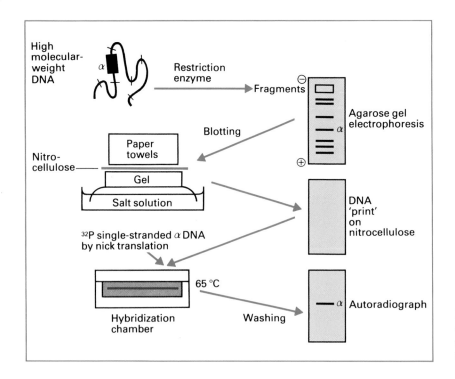

Fig. 3.7 Steps involved in the detection of an α-globin gene in a total genomic digest of human DNA using the technique of Southern blotting.

Fig. 3.8 Antenatal detection of sickle-cell genes. Normal individuals are homozygous for the β^A allele, while sufferers from sickle-cell anaemia are homozygous for the β^S allele. Heterozygous individuals have the genotype $\beta^A\beta^S$. In sickle-cell anaemia, the sixth amino acid of β-globin is changed from glutamate to valine. (A) Location of recognition sequences for restriction endonuclease *Mst* II in and around the β-globin gene. The change of A→T in codon 6 of the β-globin gene destroys the recognition site (shaded) for *Mst* II as indicated by the asterisk. (B) Electrophoretic separation of *Mst* II-generated fragments of human control DNAs (AA, AS, SS) and DNA from amniocytes (Amn). After Southern blotting and probing with a cloned β-globin gene, the normal gene and the sickle gene can be clearly distinguished. Examination of the pattern for the amniocyte DNA indicates that the fetus has the genotype $\beta^A\beta^S$, i.e. it is heterozygous.

conditions can be found in which the 'normal' probe will give a signal only with the normal gene sequence and not the mutant sequence, and vice versa. So far, this method has been used successfully to analyse the genes from patients with sickle-cell anaemia and deficiencies in α_1-antitrypsin (Fig. 3.9). This method should be applicable to any disease that is caused by a common single-base mutation and is limited in sensitivity only by the ease of detection of the gene probe.

Many individuals have ethical concerns, particularly during the second trimester, about the destruction of a fetus shown to be phenotypically positive for a genetic defect. Much more acceptable would be the diagnosis of genetic defects in preimplantation embryos which would allow those unaffected to be identified and transferred to the uterus. The first established pregnancies using this procedure have been reported in two couples known to be at risk of transmitting X-linked disorders. Two female embryos were transferred after *in vitro* fertilization, biopsy of a single cell at the six- to eight-cell stage, and sexing by DNA amplification of a Y chromosome-specific repeat sequence.

Example: Molecular profiling (fingerprinting) [8, 9, 10]. It has long been the ambition of the forensic scientist to be able to identify the origin of blood and body-fluid stains with the same degree of certainty as finger-prints. By careful use of DNA probes and Southern blotting, Alec Jeffreys and his colleagues have provided the necessary method. The method can also be used to settle cases of doubtful paternity.

Dispersed throughout the human genome are simple, tandem-repetitive regions of DNA known as *minisatellites*. These minisatellites show substantial length polymorphism which probably arises from unequal genetic exchanges. The repeat elements in a particular group of human minisatellites share a common 10–15 base-pair core:

$$\text{A}$$
$$\text{GGAGGTGGGCAGGAGG.}$$

A hybridization probe consisting of the core sequence repeated in tandem can detect many highly polymorphic minisatellites simultaneously. By using a series of variant core probes (Fig. 3.10), it is possible to detect a great number of hypervariable minisatellites, thus producing DNA profiles which are completely

M-specific probe
(C TTT CTC GTC GAT GGT CAG)

Z-specific probe
(C TTT CTT GTC GAT GGT CAG)

Fig. 3.9 Schematic representation of the use of oligonucleotide probes to detect the normal α_1-antitrypsin gene (M) and its Z variant. Human DNA obtained from normal (MM), heterozygous (MZ) and homozygous (ZZ) variant subjects is digested with a restriction endonuclease, electrophoresed and fragments Southern blotted on to a nitrocellulose membrane. The patterns shown were obtained on autoradiography of the filter following hybridization with either the normal (M-specific) or variant (Z-specific) probe.

Core sequence	GGAGGTGGGCAGGAGG A
Probe 33.6	$[(AGGGCTGGAGG)_3]_{18}$
Probe 33.15	$(AGAGGTGGGCAGGTGG)_{29}$
Probe 33.5	C $(GGGAGTGGGCAGGAGG)_{14}$

Fig. 3.10 Multi-locus probes used for DNA profiling.

specific to an individual (Fig. 3.11). Such profiles can be applied forensically (see below). Furthermore, suitable high-molecular-weight DNA can be isolated from such stains as blood and semen made on clothing several years previously and, of particular value in rape cases, sperm nuclei can be separated from the vaginal cellular debris present in semen-contaminated vaginal swabs.

The potential of DNA profiling is best illustrated by the following example. A Ghanaian boy was initially refused entry into Britain because the immigration

Fig. 3.11 DNA profiles of a Ghanaian family involved in an immigration dispute. Profiles of blood DNA are shown for the mother (M), the boy in dispute (X); his brother (B), sisters (S1, S2) and an unrelated individual (U). Fragments present in the mother's (M) DNA are indicated by a short horizontal line (to the right of each profile); paternal fragments absent from M but present in at least one of the undisputed siblings (B, S1, S2) are marked with a long line. Maternal and paternal fragments transmitted to X are shown with a dot.

The left-hand profile was obtained with probe 33.15 and the right-hand profile with probe 33.6 (Photo courtesy of Dr A. Jeffreys and the editor of *Nature*.)

authorities were not satisfied that the woman claiming him as her son was in fact his mother. Analysis of serum proteins and erythrocyte antigens and enzymes showed that the alleged mother and son were related but could not determine whether the woman was the boy's mother or aunt. To complicate matters, the father was not available for analysis nor was the mother certain of the boy's paternity. DNA profiles from blood samples taken from the mother and three children who were undisputedly hers as well as the alleged son were prepared by Southern blot hybridization to two of the minisatellite probes shown in Fig. 3.10. Although the father was absent, most of his DNA profile could be reconstructed from paternal-specific DNA fragments present in at least one of the three undisputed siblings but absent from the mother (Fig. 3.11). The DNA profile of the alleged son contained 61 scorable fragments, all of which were present in the mother and/or at least one of the siblings. Analysis of the data showed the following:

1 The probability that either the mother or the father by chance possesses all 61 of the alleged son's bands is 7×10^{-22}. Clearly the alleged son is part of the family.

2 There were 25 maternal-specific fragments in the 61 identified in the alleged son and the chance probability of this is 2×10^{-15}. Thus the mother and alleged son are related.

3 If the alleged mother of the boy in question is in fact a maternal aunt, the chance of her sharing the 25 maternal-specific fragments with her sister is 6×10^{-6}. When presented with these data, as well as results from conventional marker analysis, the immigration authorities allowed the boy residence in Britain.

Single-locus probes are a new generation of DNA probes which examine DNA at just one single location and reveal a pattern of just two bands in any one individual. These two bands are inherited directly from one's parents, one from the mother and one from the father. This may be used to determine paternity and examine claimed family relationships just as with the multi-locus probes described above. However, it is in forensic cases, where frequently only small evidential samples may be available, that these new probes have the greatest advantage over their multi-locus counterparts. Single-locus probes are some 50 times more sensitive and will allow a DNA profile to be made from as little as 10 ng of DNA. A bloodstain or semen stain the size of a pinhead, or a single hair root, may be all that is needed to make a positive identification.

Another major benefit of single-locus probes is their ability to facilitate analysis of a mixture of biological samples. It is particularly relevant in the investigation of multiple rape where a mixture of semen is likely to be the prime forensic evidence. Because the probes produce a pattern of just two bands per person, a six-band profile obtained from such a mixture would indicate the involvement of three men. Despite the apparent power of DNA profiling its acceptance by courts of law has not been straightforward. The legal objections to the technology are set out in Chapter 14 (p. 184).

DNA profiling is also showing itself to be of value in animal breeding. The Kennel Club of Great Britain have shown much interest in the technique and accept its value in determining the true sire of puppies, whose value consequently is raised.

WESTERN BLOTTING [11]

The Western blotting technique involves the transfer of electrophoresed protein bands from a polyacrylamide gel on to a membrane. The bound proteins then are available for analysis by a variety of specific protein—ligand interactions. Most commonly, antibodies are used to detect specific antigens although lectins have been used to identify glycoproteins. Although widely used by molecular biologists the technique is still in its infancy as far as diagnosis goes. However, now that a very robust and sensitive non-radioactive detection system is available, based on chemiluminescence, the method should be adopted by the clinical biochemist. One example of its use concerns the analysis of myocarditis in humans. Using Western blotting it has been shown that over 75% of patients with myocarditis, but only 4% of those with myocardial infarcts (heart attacks), have auto-antibodies to a major heart protein.

A VIRUS-BASED DIAGNOSTIC PROBE [12]

As indicated earlier, one of the disadvantages of nucleic acid probes is that they can readily detect the presence of microbial nucleic acids but cannot determine if the source was alive or dead. This is of particular concern to the food industry, which wants rapid methods for the detection and enumeration of *viable* microbes in food which may have been heat-treated. One such method makes use of the fact that bacteriophages can only replicate in viable bacteria. The bacteriophage

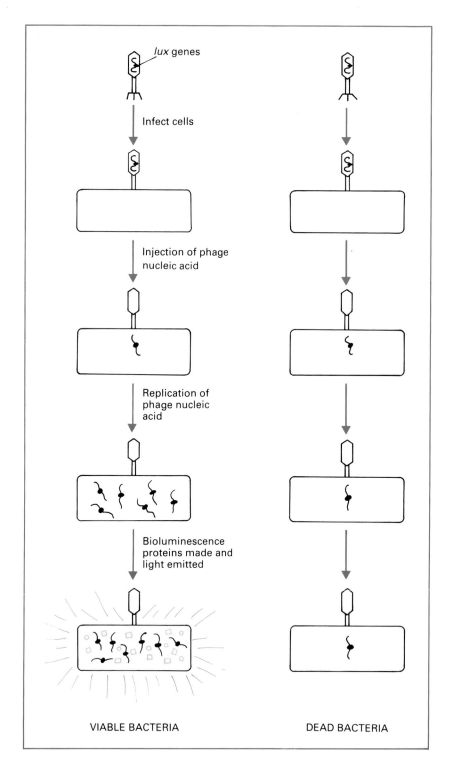

Fig. 3.12 The detection of viable bacteria using bacteriophages genetically-engineered to emit light following infection of susceptible cells. See text for details.

genome is modified by the inclusion of *lux* genes encoding bioluminescence. When such bacteriophages infect cells, the *lux* genes are expressed and light is emitted (Fig. 3.12). The light ouput can be measured in a simple luminometer and is proportional to the number of bacteria infected. It should be noted that, since multiple copies of the phage genome are present inside infected cells, there is a built-in amplification system.

Further reading

GENERAL REFERENCES

Weatherall D.J. (1985) *The New Genetics and Clinical Practice*, 2nd edn. Oxford University Press, Oxford.

CITED REFERENCES

1 Old R.W. & Primrose S.B. (1990) *Principles of Gene Manipulation: an Introduction to Genetic Engineering*, 4th edn, chapter 15. Blackwell Scientific Publications, Oxford.
2 Harris P.S. & James C.M. (1991) Use of a DNA probe for a viroid in plant pathology. In *Genetic Manipulation: Techniques and Applications* (Ed. by J.M. Grange, A. Fox & N.L. Morgan), pp. 217–225. Blackwell Scientific Publications, Oxford.
3 Durrant I., Bengel C.A., Sturrock C., Devenish A.T., Howe R., Roe S., Moore M., Scozzafava G., Proudfoot L.M.F., Richardson T.C. & McFarthing K.G. (1990) The application of enhanced chemiluminescence to membrane-based nucleic acid detection. *Biotechniques* 8, 564–70.
4 Handyside A.H., Kontogianni E.H., Hardy K. & Winston R.M.L. (1990) Pregnancies from biopsied human preimplantation embryos sexed by Y-specific DNA amplification. *Nature* 344, 768–70.
5 Pemberton J. & Amos B. (1990) DNA fingerprinting: a new dimension. *Trends in Genetics* 6, 101–3.
6 Neufield P.J. & Colman N. (1990) When science takes the witness stand. *Scientific American* 262, 18–25.
7 Uitterlinden A.G. & Vijg J. (1989) Two-dimensional DNA typing. *Trends in Biotechnology* 7, 336–41.
8 Knight P. (1990) Biosleuthing with DNA identification. *Bio/technology* 8, 505–8.
9 Lizardi P.M. & Kramer F.R. (1991) Exponential amplification of nucleic acids: new diagnostics using DNA polymerases and RNA replicases. *Trends in Biotechnology* 9, 53–8.
10 Parkin B.H. (1991) Forensic applications of DNA profiling. In *Genetic Manipulation: Techniques and Applications* (Ed. by J.M. Grange, A. Fox & N.L. Morgan), pp. 169–179. Blackwell Scientific Publications, Oxford.
11 Gershoni J.M. & Palade G.E. (1983) Protein blotting: principles and applications. *Analytical Biochemistry* 131, 1–15.
12 Stewart G., Smith T. & Denyer S. (1990) Genetic engineering for bioluminescent bacteria. *Food Science and Technology Today* 3, 19–22.

4 Protein Engineering

Many of the traditional fermentation processes for the manufacture of antibiotics, enzymes and amino acids do not utilize wild-type microorganisms because the yield of product is too low. Rather, the producing organism is subjected to successive rounds of mutagenesis until high-yielding strains are isolated. The improvement in yield is brought about by changes in proteins but what changes have been made, or even what proteins have been affected, is seldom known. As molecular biologists have learnt more about how mutations arise and how damage to DNA is repaired, the efficiency of random mutagenesis protocols has been improved. Similarly, as geneticists probing gene–protein relationships have devised clever selection procedures for rare mutants, their techniques and tricks have been adopted by those involved in strain selection. Despite the technical improvements which have been made the shotgun mutagenesis approach is limited in what it can achieve.

One of the spin-offs from recombinant DNA technology has been the development of techniques for *directed mutagenesis* of genes. There are two basic techniques. In *site-directed mutagenesis* an oligonucleotide consisting of a defined mutant sequence is hybridized to its complementary sequence in a clone of wild-type DNA. The oligonucleotide serves as a primer for *in vitro* enzymatic DNA synthesis of regions that are to remain genotypically wild-type. A double-stranded heteroduplex is formed, which subsequently is segregated *in vivo* into separate mutant and wild-type clones. In this way one or more changes can be introduced into a particular region of a gene. An alternative approach has been to use chemical synthesis of oligonucleotides to generate genes with specific mutations. In addition, a simple construct can be produced that allows for the replacement of segments by a number of mutated oligonucleotide 'cartridges' positioned between restriction endonuclease sites. The advantage of the gene synthesis approach is that it permits not only the introduction of point mutations (*micro*-modifications) but also the production of proteins in which entire polypeptide

sequences have been precisely added, deleted or replaced (*macro*-modifications).

This new-found ability to alter the amino acid sequence of a protein at will means that it is not unreasonable to consider improving proteins to meet a specific requirement (Table 4.1). This improvement is particularly relevant if it is planned to use an enzyme in an industrial process where the desired substrate or product is somewhat different from the physiological one or where the chemical conditions for the reaction are decidedly non-physiological. It also is possible to generate completely new enzymes. Using identical techniques, variants of clinically useful proteins have been isolated which have improved therapeutic properties. However, such *protein engineering* is much more complex than it might appear at first sight. Before rational changes can be made in a protein it is extremely useful to know its three-dimensional structure and what residues are important in the maintenance of this structure. If the protein has enzymic activity, it is important to know which amino acid residues have a catalytic function. This information is lacking for most proteins. Even where such information is available it is difficult to predict what effects specific changes will produce. For this reason there is a growing interest in the use of techniques for random mutagenesis of genes. Despite the fact that there is still a large degree of empiricism in protein engineering, a number of successes have been achieved.

Table 4.1 Potentially useful modifications to proteins.

Aspect	Change of interest
Stability	Increased thermostability
	Increased stability at low/high pH values
	Resistance to oxidative inactivation
Kinetics	Increased maximum velocity
	Increased affinity for substrate
	Altered substrate specificity
	Resistance to substrate/product inhibition
Biology	Altered spectrum of activity
	Altered substrate specificity

M13 vectors. As will become apparent shortly, the commonly used methods of site-directed mutagenesis require the use of a single-stranded DNA template. The easiest way to isolate such templates is to use a novel set of vectors not described in Chapter 2. These vectors are derived from a rod-shaped bacteriophage, M13, which contains a single-stranded DNA genome within a coat of viral protein. When M13 infects a susceptible cell the single-stranded viral DNA, or plus strand as it is often known, is converted to a double-stranded replicative form (RF). Initially the RF is amplified in a semi-conservative mode but replication later becomes asymmetric such that only plus strands are synthesized (Fig. 4.1). These plus strands are packaged into phage particles from which they are readily isolated and, once isolated, are ideal for site-directed mutagenesis.

To convert M13 into a cloning vector it is necessary to have convenient restriction sites in a non-essential part of the genome. Although all the genes of M13 are essential for phage multiplication, there is a non-coding region (*intergenic space*) close to the origin of replication (Fig. 4.2). A series of DNA fragments carrying multiple restriction sites have been introduced into this space to generate the M13mp series of vectors. In practice the double-stranded RF form of the vector is isolated from infected cells and genes to be cloned are inserted in the usual way. Single-stranded versions of the cloned gene, ideal for site-directed mutagenesis, are obtained by extraction of the DNA from phage particles.

Basic site-directed mutagenesis. To introduce a base change into a gene cloned in M13 a synthetic oligonucleotide, optimally 15−20 bases long, is synthesized. This oligonucleotide is complementary to the viral plus strand in the region to be mutated, except for that base which is to be changed (Fig. 4.3). The oligonucleotide is hybridized to the single-stranded template and acts as a primer for complementary strand synthesis by the Klenow fragment of DNA polymerase. The heteroduplex so formed is transformed into *Escherichia coli* where it undergoes replication and eventually gives rise to progeny phage, some of which will carry the desired mutant gene.

In theory, 50% of the progeny phage should carry the desired mutation but in practice the frequency may be much less. There are many reasons for this. For example, DNA repair mechanisms within the cell may correct the mismatches: G/T, A/C, G/G and A/A mismatches are known to be corrected with high efficiency. One way of minimizing mismatch repair is to transform the heteroduplex into cells lacking the appropriate enzymes, the so-called *mutator strains*.

Many different strategies have been developed to increase the frequency with which mutants are isolated. Most of them rely on biological selection of the mutant strand *in vivo* and often require the use of special vectors and host strains. A much more successful method is to carry out mutant strand selection *in*

Free phage containing single-stranded DNA

Phage attachment to cell

Phage DNA (plus strand)

RF formation (synthesis of minus strand)

RF replication

Asymmetric synthesis to yield plus strands only

New phage particles

Fig. 4.1 Schematic representation of the life-cycle of bacteriophage M13, which contains a single-stranded circular DNA genome. The double-stranded DNA molecule is known as a replicative form (RF).

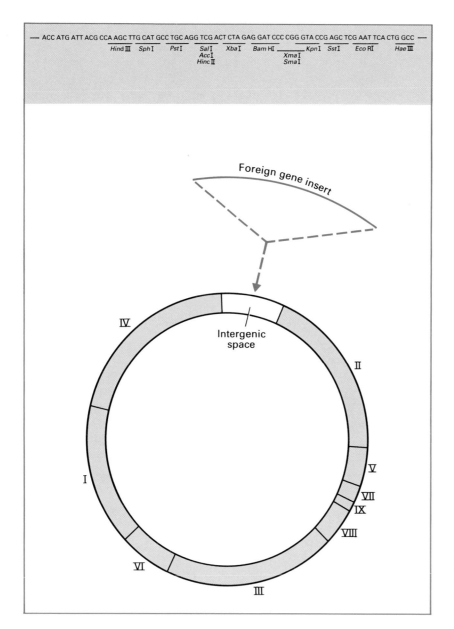

— ACC ATG ATT ACG CCA AGC TTG CAT GCC TGC AGG TCG ACT CTA GAG GAT CCC CGG GTA CCG AGC TCG AAT TCA CTG GCC —

Hind III	*Sph* I	*Pst* I	*Sal* I	*Xba* I	*Bam* HI		*Kpn* I	*Sst* I	*Eco* RI	*Hae* III
			Acc I			*Xma* I				
			Hinc II			*Sma* I				

Foreign gene insert

Intergenic
space

IV

II

I

V

VII
IX

VIII

VI

III

Fig. 4.2 The genetic map of bacteriophage M13 showing the location of the intergenic space used for gene cloning. The box shows a typical sequence which might be inserted.

vitro prior to transformation (Fig. 4.4). As before, the mutagenic oligonucleotide is annealed to the single-stranded template and extended by Klenow polymerase in the presence of DNA ligase to generate a mutant heteroduplex. If this step is carried out in the presence of a thionucleotide, the phosphorothioate DNA which is produced is resistant to cleavage with certain restriction enzymes. As a consequence, single-strand nicks are generated in DNA containing one phosphorothioate and one non-phosphorothioate strand. Such nicks present sites for exonuclease attack, permitting digestion of all or part of the non-mutant strand. The mutant strand which remains then is used as a template to reconstruct a double-stranded homo-duplex mutant molecule *in vitro*.

An alternative method of strand selection *in vitro* is the overlap extension technique which makes use of the polymerase chain reaction. Mutagenesis by this method involves the generation of DNA fragments, which, by virtue of having incorporated complementary oligo primers in independent PCR reactions, can be effectively 'fused' anywhere along the gene sequence by combining them in a second primer extension reaction. The method is illustrated in Fig. 4.5.

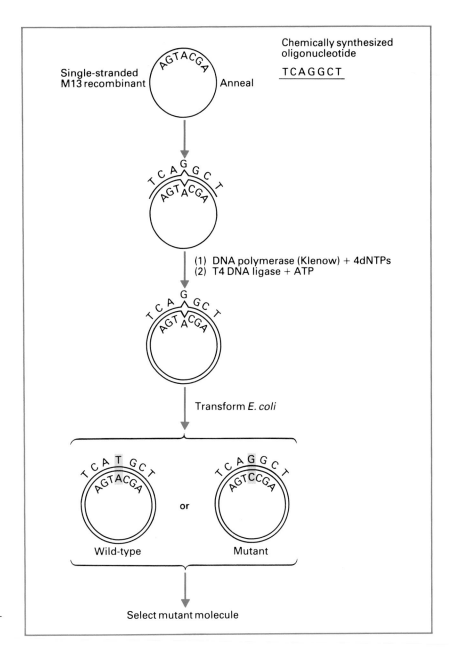

Fig. 4.3 The principle of oligonucleotide-directed mutagenesis.

In separate PCRs two fragments of the target gene sequence are amplified. Each reaction uses one flanking primer that hybridizes at one end of the target sequence (primer 'a' or 'd' in Fig. 4.5) and one internal primer that hybridizes at the site of the mutation and contains the mismatched bases (primer 'b' or 'c' in Fig. 4.5). Since the product generated in a PCR incorporates the primers, the wild-type sequence will not be amplified. By using two internal primers that overlap, the two fragments AB and CD, generated in the first PCR, can be fused by denaturing and annealing them in a subsequent primer extension reaction. The overlap allows one strand from each fragment to act as a primer on the other, and extension of this overlap results in the mutant product. Even though the annealing of the short overlap between the two fragments may occur at low frequency, the inclusion of additional flanking primers ('a' and 'd' in Fig. 4.5) allows the 'fusion' product that is formed to be amplified by PCR.

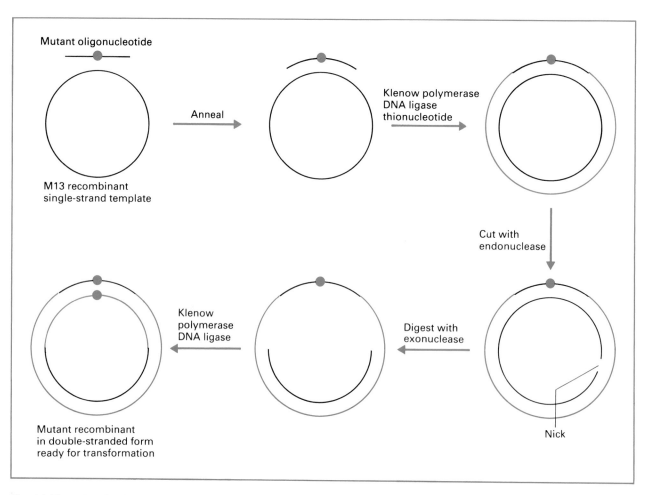

Fig. 4.4 The principle of thionucleotide selection of mutant DNA molecules (see text for details). The mutant base is shown as a red spot. DNA synthesized in the presence of thionucleotides is shown by a red line.

Multiple mutations. Site-directed mutagenesis need not be restricted to single base changes. By appropriate choice of oligonucleotides it is possible to introduce multiple-point mutations, insertions and deletions (Fig. 4.6). Thus by using a 30-base oligonucleotide (referred to as a 30-mer, meaning a polymer of 30 nucleotides) 11 base changes were introduced into a single region of a gene. Multiple base changes at the *same* position can be made in a single experiment by using mixed oligonucleotide primers. Large deletions can be made between two sites if one half of the primer is complementary to the first site and the other half complementary to the second site. In this way an 18-mer has been used to delete over 1100 nucleotides. Insertions can be made likewise, although the length of the insert is limited by the length of the primer. Using normal-length oligonucleotides (e.g. less than 40-mers) the largest insert which can be achieved is 10−15 bases.

INCREASED PROTEIN STABILITY

The goal of many protein engineering programmes is to achieve increased protein stability. This may be either increased half-life or thermostability of an enzyme used in an enzyme reactor, increased shelf-life of a protein for therapeutic use or resistance to inactivation resulting from oxidation of important amino acid residues. These goals have been achieved as the examples given below show. Increased thermostability is particularly important since there are benefits from operating enzyme reactors at elevated temperatures. For example, reaction times are shorter and there is less risk of microbial spoilage of the enzyme.

Fig. 4.5 Schematic diagram of site-directed mutagenesis by overlap extension. The ds DNA and synthetic oligos are represented by lines with arrows indicating the 5′-to-3′ orientation. The site of mutagenesis is indicated by the small black rectangle. Oligos are denoted by lower-case letters and PCR products are denoted by pairs of upper-case letters corresponding to the oligo primers used to generate that product. The boxed portion of the figure represents the proposed intermediate steps taking place during the course of reaction (3), where the denatured fragments anneal at the overlap and are extended 3′ by DNA polymerase (dotted line) to form the mutant fusion product. By adding additional primers 'a' and 'd' the mutant fusion product is further amplified by PCR. Reproduced courtesy of Dr L.R. Pease and the editor of *Gene*.

Furthermore, cooling of reactors is not always efficient, particularly when exothermic reactions are being performed.

Example: Increased thermostability of lysozyme [4, 5, 6, 7, 8, 9, 22]. One theoretical way of increasing the thermal stability of a protein is to introduce a disulphide bond where none already exists in the molecule. Although not of commercial importance, T4 lysozyme is the only enzyme in which this has been achieved. In T4 lysozyme there are two cysteine residues (54 and 97) which are unpaired. Careful examination of the three-dimensional structure indicated that conversion of isoleucine residue 3 to a cysteine should enable a disulphide bridge to form with cysteine 97. Site-directed mutagenesis was used to convert the ATA isoleucine codon in a cloned T4 lysozyme gene to a TGT cysteine

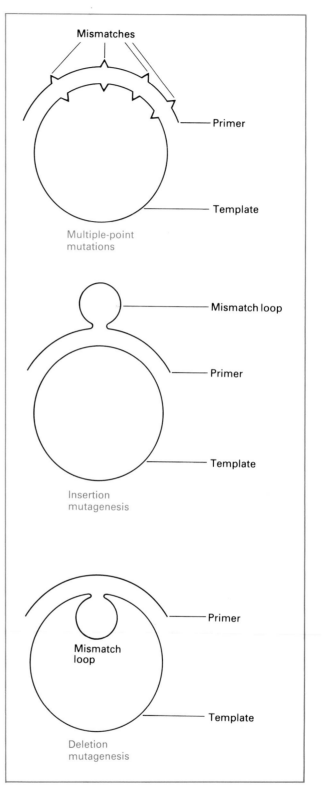

Fig. 4.6 The use of oligonucleotide-directed mutagenesis to generate multiple-point mutations, insertions and deletions.

codon. Biochemical studies on the mutant enzyme confirmed that a cross-link did form as expected. Under oxidizing conditions the mutant and wild-type enzymes had the same enzymic activity but the mutant had greater thermostability. At 67 °C the half-life of the wild-type enzyme was 11 min, whereas that of the mutant enzyme was 28 min. Furthermore, the activity of the mutant enzyme never fell below 50% of maximum, whereas the wild-type enzyme's activity dropped to 0.2% after 180 min.

Despite the success achieved with lysozyme, the introduction of disulphide bridges does not appear to be a generic method of increasing protein thermostability. Other kinds of amino acid substitutions have led to generally applicable methods. For example, glycine has no β-carbon and allows more conformational freedom in a polypeptide chain than alanine. Thus, in a variety of proteins, changing glycine residues to alanine residues has led to increased protein stability, probably by reducing the entropy loss on unfolding. Similar effects have been achieved by introducing proline residues. Another way to increase thermostability is to increase the number of the stabilizing interactions in the folded protein. In the case of glutamine synthetase, increasing the internal hydrophobicity by replacement of a glutamine residue resulted in stabilization of the protein.

Irreversible inactivation of some proteins is triggered by deamination of asparagine residues, by intramolecular thio/disulphide exchange, or by destruction of methionine residues. Substitutes for asparagine, cysteine and methionine, have been shown to prevent such inactivation and to increase thermostability.

Several studies have shown that the effects of amino acid substitutions may be localized so that their effects on the stability are additive. For example, six amino acid substitutions have been made in subtilisin, each of which increases the stability by $0.3-1.3\,\text{kcal}\,\text{mol}^{-1}$. When they are all incorporated the stability increases by $3.8\,\text{kcal}\,\text{mol}^{-1}$. Thus, it seems likely that it will be possible to construct very stable enzymes when they are needed and this should increase their usefulness in many biotechnological applications.

Example: Enhanced specific activity of recombinant interferon-β [10]. A different kind of stability has been achieved with human interferon-β, a protein with three cysteine residues. When interferon-β is synthesized in *E. coli* it forms aggregates and the specific activity of the protein is 10^6 antiviral units/mg, one hundred times less than fibroblast-derived material. One possible explanation for the low specific activity of the recombinant material is incorrect disulphide bond formation. Thus when cysteine residue 17, which should be present as a free thiol, was changed to serine the specific activity of the mutant interferon was 10^8 units/mg, i.e. it had full specific activity. An additional benefit of this cysteine to serine change was increased stability on prolonged storage, an important feature of any protein which is to be sold for therapeutic use.

Example: Improved α₁-antitrypsin [11]. α₁-Antitrypsin (AAT) is the archetypal member of a family of homologous proteins known as serine protease inhibitors or serpins. *In vivo* the function of AAT is the inhibition of neutrophil elastase. Inhibition occurs by a rapid association between the two proteins followed by cleavage of the AAT between methionine residue 358 and serine residue 359 (Fig. 4.7). After cleavage there is negligible dissociation of the complex.

Oxidative stress, as occurs in a number of pathological conditions (see Box. 4·1), results in the con-

---

Box 4·1 α₁-Antitrypsin, emphysema and smoking

Cumulative damage to lung tissue by neutrophil elastase is thought to be responsible for the development of emphysema, an irreversible lung disease characterized by loss of lung elasticity. The primary defence against elastase damage is α₁-antitrypsin (AAT), a glycosylated serum protein of 394 amino acids. The function of AAT is known because its genetic deficiency leads to a premature breakdown of connective tissue. In healthy individuals the AAT diffuses into tissue spaces where it complexes with neutrophil elastase and these complexes are subsequently removed from the circulation and catabolized in the liver and spleen.

Smokers are more prone to emphysema because smoking results in an increased concentration of leucocytes in the lung and consequently increased exposure to neutrophil elastase. In addition, leucocytes liberate oxygen-free radicals and these can oxidize the critical methionine residue 358 (see main text) rendering the AAT ineffective. Exposure of AAT to oxidants can also occur with ozone pollution or therapeutic use of pure oxygen.

Fig. 4.7 The cleavage of α₁-antitrypsin on binding to neutrophil elastase.

version of methionine 358 to methionine sulphoxide. Since methionine sulphoxide is much bulkier than methionine it does not fit into the active site of elastase and hence oxidized AAT is a poor inhibitor. By means of site-directed mutagenesis, an oxidation-resistant mutant of AAT has been constructed by replacing methionine 358 with valine. In a laboratory model of inflammation, the degradation of basement membrane collagen by stimulated neutrophils, the modified AAT was an effective inhibitor of elastase and was not inactivated by oxidation. Clinically this could be important since intravenous replacement therapy with plasma concentrates of AAT is already being tested on patients with a genetic deficiency in AAT production.

ALTERING THE KINETIC PROPERTIES OF ENZYMES

Many enzymes are used in large-scale industrial processes and a number of examples are given in later chapters (see Tables 6.8 and 7.3). Effective use of these enzymes requires a knowledge of enzyme kinetics (see Box 4·2). For most applications the enzymes are used under conditions where there is a vast excess of substrate and v_{max} is the only kinetic parameter of importance. It should be noted that v_{max} is usually measured during the initial stages of the reaction. As the reaction proceeds the concentration of P will increase and the rate constant k_4 (see Box 4·2) can no longer be ignored unless the reaction is irreversible. High concentrations of product can also inhibit the enzyme, leading to a decrease in the reaction velocity.

If an enzyme is being used to *destroy* a substrate present at low concentrations, the K_m of the enzyme is the key parameter. The best example is provided by asparaginase. In certain leukaemias the growth of the malignant white blood cells is dependent upon the presence in the bloodstream of asparagine. Thus intra-

venous administration of the enzyme asparaginase is therapeutically beneficial since it cleaves asparagine to aspartate. During a search for sources of asparaginase the puzzling discovery was made that not all asparaginases are effective in suppressing leukaemia. The explanation was eventually provided by kinetic analysis; the enzymes from the different sources differ widely in their K_m for asparagine. The administration of asparaginase will be therapeutically useful only if its K_m value is low enough to hydrolyse asparagine rapidly at the low concentration at which it is present in blood.

Example: Altering the K_{cat} *and* K_m *of subtilisin* [12]. The industrial applications of many enzymes are restricted because of oxidative inactivation in an analogous manner to that of AAT. This is especially true for enzymes containing methionine, cysteine or tryptophan residues in or around the active site. Like AAT, the protease subtilisin has a methionine residue in its active site and, being a component of biological detergents, subtilisin is exposed to oxidative stress during use! Using site-directed mutagenesis, variant enzymes have been obtained in which this methionine residue is replaced by any one of the other 19 natural amino acids. The different mutant enzymes were found to vary widely in activity (see Table 4.2 for representative examples). With the exception of the methionine to cysteine replacement, all had decreased activity. Mutants containing non-oxidizable amino acids, e.g. serine, alanine or leucine, were resistant to inactivation by 1 mol/l hydrogen peroxide, whereas the methionine and cysteine enzymes were rapidly inactivated (Fig. 4.8).

Evaluating this kinetic information in terms of the usefulness of the subtilisin mutants for inclusion in detergents is a little complex. With the exception of the cysteine replacement, all the mutants have a decreased turnover number and, at first sight, would appear to be less useful but that is in the absence of oxidative

Table 4.2 Kinetic constants for selected mutants of subtilisin in which methionine residue 222 is replaced by other amino acids.

Amino acid 222	K_{cat} (per s)	K_m (mol/l)
Methionine	50	1.4×10^{-4}
Cysteine	84	4.8×10^{-4}
Serine	27	6.3×10^{-4}
Alanine	40	7.3×10^{-4}
Leucine	5	2.6×10^{-4}

Box 4·2 Basic enzyme kinetics

Most enzymatic reactions use more than one substrate in concert but it is easier to explain the basic principles of enzyme kinetics by considering a one-substrate reaction. Such a reaction can be summarized as:

Enzyme (E) + ⇌ enzyme − ⇌ enzyme +
substrate (S)   substrate   products (P)
                   complex (ES)

From this it should be apparent that increasing the substrate concentration at a fixed level of enzyme increases the reaction velocity (Diagram 1). Maximum velocity (v_{max}) is achieved when all the enzyme is in the form of the ES complex.

For the purposes of kinetic analysis it is convenient to examine that period of the enzymatic reaction known as the *steady state*. This occurs shortly after mixing enzyme and substrate and constitutes the time interval when the rate of reaction is a linear function of time (Diagram 2). Usually substrate is present in much higher molar concentrations than enzyme and, since the initial period of the reaction is being examined, the free substrate concentration is approximately equal to the total substrate added to the reaction mixture.

There are four rate constants associated with the above reaction sequence:

Diagram 1 Effect of substrate concentration on the rate of an enzyme-catalysed reaction.

$$E + S \underset{k_3}{\overset{k_1}{\rightleftharpoons}} ES \underset{k_4}{\overset{k_2}{\rightleftharpoons}} E + P$$

If only the *initial* rate of reaction is being considered, the rate constant k_4 can be ignored because not enough product will be present to make the reverse reaction proceed at a significant rate. If v is the rate of product formation then

$$v = k_2 \, [E]_t \qquad (1)$$

Fig. 4.8 Resistance to oxidation by subtilisin mutants in which methionine residue 222 is replaced by alanine or cysteine. Activity remaining after exposure to 0.1 mol/l (a) or 1.0 mol/l (b) hydrogen peroxide.

Box 4·2 (*continued*)

Diagram 2 Time-course of the formation of product and enzyme–substrate complex in an enzyme-catalysed reaction.

and since the maximum initial rate v_{max} should be obtained when all the enzyme is in the form of the ES complex,

$$v_{max} = k_2 [E]_t \qquad (2)$$

where $[E]_t$ represents the total enzyme concentration, i.e. the sum of the free and combined enzyme. The kinetic constant k_2 is referred to as the *turnover number* and sometimes is symbolized by K_{cat}. From equation (2) it can be seen that the turnover number is equal to the number of substrate molecules converted into product in a given time (usually 1 s) by a mole of enzyme when the enzyme is fully complexed with substrate. For most enzymes this number falls in the range 10^1-10^4/s.

Under steady-state conditions the rate of formation of the ES complex is equal to the rate of its breakdown. This rate is known as K_m or the *Michaelis constant* and is represented by the expression:

$$K_m = \frac{k_2 + k_3}{k_1} \qquad (3)$$

Numerically K_m is equal to the concentration (expressed in mol/l) of the substrate which gives half the maximum velocity. For any given enzyme the value of K_m depends on the substrate and the conditions of the reaction such as pH, temperature, etc. For most enzymes the value of K_m lies between 10^{-1} and 10^{-6} mol/l. The lower the value of K_m, the stronger the complex between enzyme and substrate.

stress. If oxidative stress during use approximates to 1 mol/l hydrogen peroxide, then clearly the alanine-containing variant would be most useful (Fig. 4.8). If the stress approximates to 0.1 mol/l hydrogen peroxide, then perhaps the cysteine-containing variant would be more appropriate. However, there is another consideration. Unlike many enzymes of commercial importance, subtilisin in detergents operates at low substrate concentration. Thus the affinity of the enzyme for its substrate, i.e. the K_m value, cannot be ignored and the replacement of methionine by other amino acids leads to a decreased affinity.

Example: Altering the pH dependence of enzyme catalysis [12]. One prospect for protein engineering is the tailoring of the pH dependence of enzyme catalysis to optimize activity in industrial processes. Chemical studies have shown that the pH dependence of catalysis by serine proteases alters with changes in overall surface charge. Thus a possible way of modifying pH dependence would be to alter the electrostatic environment of the active site by protein engineering and so change the pK_a values of ionic catalytic groups. That such alterations are possible has been shown again with subtilisin where a single amino acid change, aspartate to serine, reduced the pH optimum by 0.3 units.

SUCCESSFUL MUTAGENESIS WITHOUT KNOWLEDGE OF PROTEIN TERTIARY STRUCTURE [13, 14]

Successful protein engineering of the kind discussed above depends upon the availability of reliable structural information. At present, X-ray diffraction methods are the only techniques that can provide the required data although two-dimensional nuclear magnetic resonance (NMR) may be an alternative procedure in the future. Protein crystallography is a very laborious process despite major advances in the collection and analysis of diffraction data. The most unpredictable step, and hence the one which is rate-limiting, is the preparation of protein crystals of the correct quality. Indeed, some workers have been forced to make use of the zero gravity aboard space shuttles to eliminate convective effects and improve crystallization! A recent finding that the initiation and rate of crystallization are significantly enhanced at high pressures may go a long way towards facilitating the isolation of good crystals.

The use of sophisticated computer graphics has eliminated the need for building physical models. The protein structure is simply fitted to the electron density map provided by X-ray diffraction analysis. In addition, the use of computers facilitates the estimation of the perturbations that would result from specific amino acid substitutions of a known protein structure.

When the tertiary structure is not available successful modification through site-directed mutagenesis may be achieved if a homologous protein of known structure is available. Using this strategy, a neutral protease from *Bacillus stearothermophilus* was stabilized using thermolysin as a model protein.

Stabilization by thermal adaptation [15]. In the absence of a tertiary structure protein stabilization can be achieved by selecting for gene function at elevated temperatures. As an example, a gene encoding a nucleotidyl transferase which specifies kanamycin resistance was isolated from a mesophile and cloned in a thermophilic bacterium. The thermophile was grown at elevated temperatures and when selection was made for kanamycin resistance two naturally occurring mutants were selected. In one mutant, aspartate residue 80 was converted to tyrosine and in the other threonine residue 130 was converted to lysine. Both the substituted proteins are more thermostable than the original mesophilic protein.

Random mutagenesis [16]. This is a powerful technique for probing structure–function relationships in proteins and is particularly useful for generating desirable mutants in the absence of information on tertiary structure. Oligonucleotide-directed *in vitro* mutagenesis can be adapted for the random mutagenesis of cloned genes by the use of 'spiked' oligonucleotide primers. Preparation of such spiked oligonucleotides can be accomplished by introducing a low level of contamination with three 'wrong' phosphoramidites in each coupling cycle during automated DNA synthesis (see p. 20). These spiked oligonucleotides contain errors that are close to random in terms of the nature and positions of misincorporated nucleotides. The family of mutagenic oligonucleotides thus produced is then used in either of the procedures for directed mutagenesis shown in Figs 4.3 and 4.4.

The spiked primer approach to random mutagenesis offers a degree of flexibility and control that is absent from earlier methods. First, there are no hot spots in a synthetic oligonucleotide, and, provided that the synthesis is performed with pure reagents of equal reactivity, there is an equal probability of mutation at every position along the primer. Second, the mutagenic window is precisely determined by the choice of the oligonucleotide sequence. Third, the level of contamination can be varied to achieve any desired frequency of point mutations in the target window. Fourth, the procedure can be adapted to use pure ('unspiked') reagents at particular positions in the sequence when no base variation is desired.

INCREASING THE BIOACTIVITY OF PROTEINS

Many human proteins are being tested as potential therapeutic agents and a small number of them, produced by recombinant DNA technology, are already in the market-place (see Chapter 6). Protein engineering is now being used to generate second-generation variants with improved pharmacokinetics, structure, stability and bioavailability. One such example has already been cited, the oxidation-resistant variant of α_1-antitrypsin (see p. 48).

Example: Altering the pharmacokinetics of tissue plasminogen activator (tPA) [17]. The enzyme tPA (see Box 4·3) is used therapeutically to lyse blood clots that cause myocardial infarction and pulmonary embolism. When tPA is administered as a bolus to humans over 50% of the drug is cleared from the bloodstream in 5 minutes. To circumvent this problem of clearance, the drug has to be dosed by intravenous infusion over extended time periods. tPA is glycosylated and many plasma glycoproteins are known to be recognized

Box 4·3 Antithrombolytic and anti-clotting agents

Blood clotting is a complex process because it must be highly controlled. There are a number of diseases caused by the blockage of blood vessels by intravascular clots, i.e. *thrombosis*. Clots occurring in the coronary arteries supplying the heart cause myocardial infarcts (heart attacks) whereas those occurring in the arteries of the brain or lungs result in strokes and pulmonary embolisms respectively. Normally clots are broken down by the action of the enzyme plasmin. This enzyme is present in the inactive form, plasminogen, but is activated by a group of serine proteases called plasminogen activators (Diagram 1). One of these is urokinase, which is isolated from human urine and which activates free plasminogen. The other is tissue plasminogen activator (tPA), which is produced by a number of cell types. tPA binds to fibrin and is selective for plasminogen bound to thrombi (clots). Thus it only works when it is needed.

Although blood clot formation is an essential early step in wound healing there are many clinical situations where thrombus formation must be minimized. For example, clotting must be minimized in patients who are recovering from a thrombosis or from certain kinds of micro-

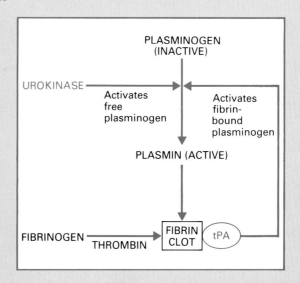

surgery. A key step in thrombus formation is the proteolytic conversion of fibrinogen to fibrin, which is mediated by thrombin. Many blood-sucking animals such as leeches and mosquitoes produce thrombin inhibitors.

---

by liver receptors specific for certain oligosaccharide linkages. One residue in the tPA molecule which is glycosylated is asparagine residue 120. When this residue was replaced with glutamine, 'longer-acting tPA' was produced which has a significantly increased circulating half-life (Fig. 4.9).

Example: Increasing the potency of hirudin [18]. Hirudin, a 65 amino acid polypeptide secreted from the leech salivary gland, is a potent thrombin inhibitor. Recombinant hirudin is being evaluated as an anti-coagulant. When asparagine residue 47 was converted to lysine or arginine, the potency in a test-tube clotting assay increased fourfold. When tested in an animal model the antithrombotic effects of the variants were over 20-fold higher than the native protein and five times higher than heparin.

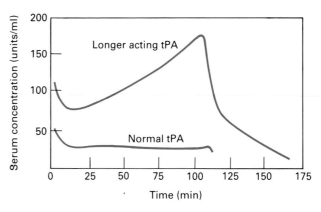

Fig. 4.9 Clearance rates of normal and 'longer-acting' tissue plasminogen activator from rabbit blood following an 11-minute infusion. Note that high levels of 'longer acting' tPA accumulated during the infusion and that more activity persisted after cessation of the infusion than was observed for normal tPA. (Figure modified from reference 17.)

The enzyme staphylococcal nuclease non-specifically hydrolyzes the phosphodiester bonds of single-stranded RNA and DNA. The three-dimensional structure of the enzyme suggested that coupling of an oligonucleotide-binding domain to lysine residue 116 or valine residue 114 on the enzyme surface should generate an enzyme with sequence specificity. The binding domain selected was an oligonucleotide since these bind to single-stranded DNA or RNA with high specificity and affinity. This oligonucleotide binding domain was synthesized by gene synthesis (see p. 20) such that there was a 3′-terminal thiol. Using site-directed mutagenesis a unique thiol was introduced into staphylococcal nuclease by replacement of lysine 116 or valine 114 with cysteine. In the presence of reducing agents the oligonucleotide coupled to the cysteine mutants of the nuclease in high yield. Hybrid enzymes produced in this way exhibited sequence-specific cleavage of target nucleic acids unlike the natural molecules.

MACRO-MODIFICATION TO PROTEINS

With site-directed mutagenesis the aim is to replace, via base changes in a gene, one, or at most a few, amino acids in a protein sequence which will in some way 'improve' the protein in question. Much larger modifications can be made to a protein; for example, part of a gene can be deleted by eliminating a restriction fragment or by chemically synthesizing only part of a gene. In this way it is possible to produce the Klenow fragment of DNA polymerase (see Box 3.1, p. 29) free of contamination with the 3′ → 5′ exonuclease activity associated with the intact enzyme. Conversely, it is possible to insert additional amino acids into a protein sequence and this has been done to create purification fusions (see below) and to improve the stability of foreign proteins synthesized in E. coli. Finally, all or part of one gene can be fused with all or part of another to generate completely novel proteins, as has been done with the interferons (see below).

Example: Novel interferons [20]. Human interferon-α and -β elicit a number of biological responses such as conferring resistance to viral infections, macrophage and killer cell activation, inhibition of cell proliferation and modulation of the immune response. When ad-

ministered clinically a reversible dose-limited toxicity is frequently observed and this could be because of the action of interferons on so many cell types. By constructing hybrids between different interferons it may be possible to produce novel interferons with superior clinical properties, e.g. increased potency or decreased toxicity. That this is feasible has been shown by the synthesis of a 'consensus' interferon-α with a 15-fold higher specific activity than any natural interferon-α. Hybrids have also been constructed in which portions of the interferon-β molecule have been replaced with corresponding segments from interferon-α. Some of these hybrid interferons have an altered antiviral or antiproliferative profile (Table 4.3).

Genetically engineered antibodies have been produced in a similar way; for example, mouse and rat antibodies have been converted into ones with far greater similarity to human antibodies. This is discussed in more detail in Chapter 10.

Example: PIXY-321. This is a fused form of granulocyte macrophage colony stimulating factor and interleukin 3. It binds to a dual receptor on bone marrow cells and stimulates both platelet and white blood cell production.

Example: Purification fusions [21]. Recombinant DNA technology can be used to minimize or eliminate some of the problems associated with protein purification, particularly when the protein is destined for therapeutic use. Many foreign proteins when produced in E. coli are particularly susceptible to protease attack

Table 4.3 Properties of some modified interferons.

Amino acid change	Antiviral activity relative to interferon-β	Antiproliferative activity relative to interferon-β
Interferon-β residues 2−7 replaced by interferon-α₂ residues 1−5 and cysteine 17 replaced by serine	2.3	4.4
Interferon-β residues 9−56 replaced by interferon-α₁ residues 7−54	0.005	1.9
Interferon-β residues 82−105 replaced by interferon-α₁ residues 80−103	0.76	3.3

Fig. 4.10 A tripartite protein fusion which can be used to facilitate purification of a protein.

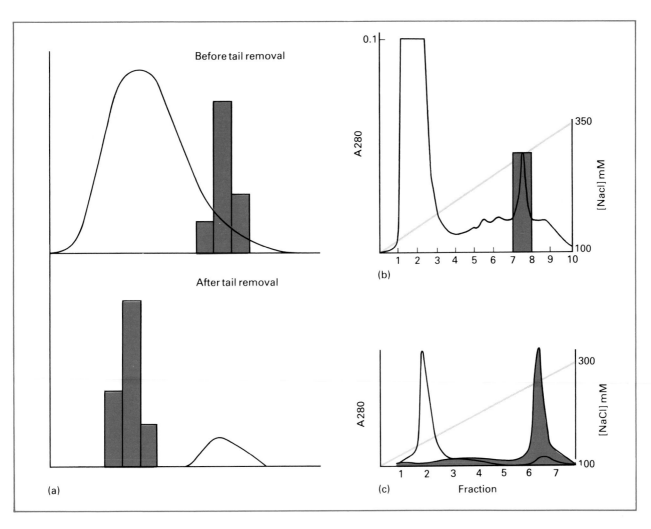

Fig. 4.11 The use of polyarginine-tailing to facilitate protein purification. (a) (*left*) Schematic representation of a hypothetical protein before and after enzymic removal of the C-terminal arginine residues. (b) (*top right*) Separation of polyarginine-tailed urogastrone from the bulk of the proteins in an *E. coli* cell extract. The red line shows the salt gradient and the tinted bar the urogastrone activity. (c) (*bottom right*) Chromatographic behaviour of tailed and untailed urogastrone. The red line shows the salt gradient.

such that little intact protein can be isolated. These labile proteins can be protected by producing them as fusions with *E. coli* proteins and such fusion proteins can facilitate purification. One generalized method which has been developed is to build a gene which encodes the target protein joined to β-galactosidase via a collagen peptide linker (Fig. 4.10). The tripartite protein is first purified by exploiting its affinity for resin-bound *p*-aminophenyl thiogalactoside residues and the target protein released by controlled digestion with collagenase. Any protein for which there is a suitable affinity purification method can be used instead of β-galactosidase, e.g. protein A-containing fusions can be purified on columns of immobilized immunoglobulin.

An alternative approach to purification fusions is polyarginine-tailing. In this method the gene sequence encoding the desired protein is extended by the inclusion at the 3' end of a number of codons for arginine. When such genes are expressed the resultant proteins have a polyarginine 'tail' which makes them more basic. Upon ion-exchange chromatography such proteins are separated from the bulk of the host cell proteins, which are more acidic (Fig. 4.11). The polyarginine tail is then removed with the enzyme carboxypeptidase B, which for convenience can be immobilized. The de-tailed protein is rechromatographed on an identical ion-exchange resin to separate it from any remaining contaminating proteins which will be more basic.

Further reading

GENERAL REFERENCES

Blundell T.L., Sibanda B.L., Sternberg M.J.E. & Thornton J.M. (1987) Knowledge based prediction of protein structures and the design of novel molecules. *Nature* **326**, 347–52.

Hilvert D. (1991) Extending the chemistry of enzymes and abzymes. *Trends in Biotechnology* **9**, 11–17.

Knowles J.R. (1987) Tinkering with enzymes: what are we learning? *Science* **236**, 1252–8.

Livingston D.J. & McPherson J.M. (1989) Protein engineering: new approaches to improved therapeutic proteins, Part II. *Pharmaceutical Technology* June, 26–34.

Old R.W. & Primrose S.B. (1990) *Principles of Gene Manipulation: an Introduction to Genetic Engineering*, 4th edn. Blackwell Scientific Publications, Oxford.

Pace C.N. (1990) Measuring and increasing protein stability. *Trends in Biotechnology* **8**, 93–8.

Wagner C.R. & Benkovic S.J. (1990) Site directed mutagenesis: a tool for enzyme mechanism dissection. *Trends in Biotechnology* **8**, 263–70.

CITED REFERENCES

1 Zoller M.J. & Smith M. (1983) Oligonucleotide-directed mutagenesis of DNA fragments cloned into M13 vectors. *Methods in Enzymology* **100**, 468–500.

2 Nakamaye K.I. & Eckstein F. (1986) Inhibition of restriction endonuclease Nci I cleavage by phosphorothioate groups and its application to oligonucleotide-directed mutagenesis. *Nucleic Acids Research* **14**, 9679–98.

3 Ho S.N., Hunt H.D., Horton R.M., Pullen J.K. & Pease L.R. (1989) Site-directed mutagenesis by overlap extension using the polymerase chain reaction. *Gene* **77**, 51–9.

4 Perry L.J. & Wetzel R. (1984) Disulfide bond engineered into T4 lysozyme: stabilization of the protein towards thermal inactivation. *Science* **226**, 555–7.

5 Matthews B.W., Nicholson H. & Becktel W.J. (1987) Enhanced protein thermostability from site-directed mutations that decrease the entropy of unfolding. *Proceedings of the National Academy of Sciences USA* **84**, 6663–7.

6 Yutani K., Ogasahara K., Tsujita T. & Sugino Y. (1987) Dependence of conformational stability on hydrophobicity of the amino acid residue in a series of variant proteins substituted at a unique position of tryptophan synthase α subunit. *Proceedings of the National Academy of Sciences USA* **84**, 4441–4.

7 Ahern T.J., Casal J.I., Petsko G.A. & Klibanov A.M. (1987) Control of oligomeric enzyme thermostability by protein engineering. *Proceedings of the National Academy of Sciences USA* **84**, 675–9.

8 Perry L.J. & Wetzel R. (1987) The role of cysteine oxidation in the thermal inactivation of T4 lysozyme. *Protein Engineering* **1**, 101–5.

9 Pantoliano M.W., Whitlow M., Wood J.F., Dodd S.W., Hardman K.D., Rollence M.L. & Bryan P.N. (1989) Large increases in general stability for subtilisin BPN' through incremental changes in the free energy of unfolding. *Biochemistry* **28**, 7205–13.

10 Mark D.F., Lu S.D., Creasey A.A., Yamamoto R. & Lin L.S. (1984) Site specific mutagenesis of the human fibroblast interferon gene. *Proceedings of the National Academy of Sciences USA* **81**, 5662–6.

11 Courtney M., Jalkt S., Terrier L.-H., Benavente A., Crystal R.G. & Lecocq J.-P. (1985) Synthesis in *E. coli* of alpha₁-antitrypsin variants of therapeutic potential for emphysema and thrombosis. *Nature* **313**, 149–51.

12 Wells J.A. & Estell D.A. (1988) Subtilisin—an enzyme designed to be engineered. *Trends in Biochemistry* **13**, 291–7.

13 Imanaka T., Shibazaki M. & Takagi M. (1986) A new way of enhancing the thermostability of proteases. *Nature* **324**, 695–7.

14 Viuri K., Kaipainen E., Kivimaki J., Niemi H., Leisda M. & Palosaari S. (1990) A new method for protein crystallization using high pressure *Bio/technology* **8**, 547–9.

15 Liao H., McKenzie T. & Hageman R. (1986) Isolation of a thermostable enzyme variant by cloning and selection in a thermophile. *Proceedings of the National Academy of Sciences USA* **83**, 576–80.

16 Hermes J.D., Parekh S.M., Blacklow S.C., Köster H. & Knowles J.R. (1989) A reliable method for random muta-

genesis: the generation of mutant libraries using spiked oligo-deoxyribonucleotide primers. *Gene* **84**, 143−51.

17 Lau D., Kuzma G., Wei C-M., Livingston D.J. & Hsiung N. (1987) A modified human tissue plasminogen activator with extended half-life *in vivo*. *Bio/technology* **5**, 953−8.

18 European Patent Application O 273 800 A2, 6 July 1988.

19 Schultz P.G. (1988) The interplay between chemistry and biology in the design of enzymatic catalysts. *Science* **240**, 426−33.

20 Weck P.K., Apperson S., Stebbing N., Gray P.W., Leung D., Shepard H.M. & Goeddel D.V. (1982) Antiviral activities of hybrids of two major human leucocyte interferons. *Nucleic Acids Research* **9**, 6153−66.

21 Sassenfeld H.M. (1990) Engineering proteins for purification. *Trends in Biotechnology* **8**, 88−93.

22 Sanderg W.S. & Terwilliger T.C. (1991) Repacking protein interiors. *Trends in Biotechnology* **9**, 59−63.

Part III
The Exploitation of Microbes

5 The Large-scale Cultivation of Microbes

Much of biotechnology centres around the large-scale cultivation (10−500 000 l) of microbes for the production of cells themselves, e.g. biomass, or the production of useful metabolites or proteins. Such large-scale cultures are often referred to as industrial fermentations. *Sensu stricto*, a fermentation is a biological process occurring in the absence of air (oxygen). The first large-scale microbial processes to be developed, brewing and acetone−butanol production, are truly fermentations. Ethanol is formed by yeast cells as a means of regenerating oxidized pyridine nucleotides in the absence of oxygen (Fig. 5.1). Acetone and butanol are formed for the same reason but in this instance the producing organism is a strict anaerobe. However, the term 'industrial fermentation' is now applied to any large-scale cultivation of microbes even though most of them are aerobic. This is ironic for oxygen supply is the single most important factor limiting the

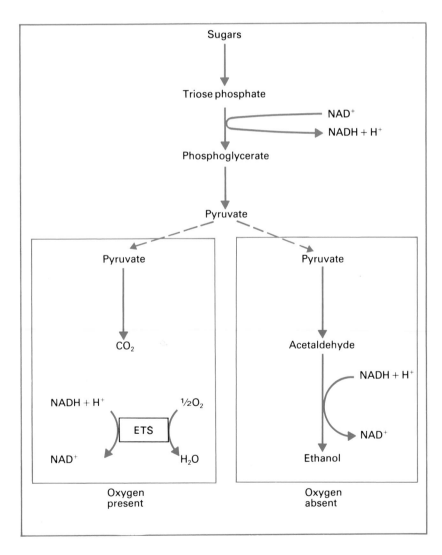

Fig. 5.1 The regeneration of oxidized pyridine nucleotides by yeast cells in the presence and absence of oxygen.

efficiency of aerobic processes and modern fermenter design and practice centres around the provision of adequate amounts of oxygen! Obligate aerobes require oxygen for growth: limit the supply of oxygen and growth ceases. Facultative anaerobes can grow in the absence of oxygen but growth rates are lower, leading to longer process times, and cell yields are greatly reduced with a concomitant reduction in product levels.

THE PROBLEM OF OXYGEN SUPPLY

A rapidly growing culture has a very high demand for oxygen. More correctly, it has a high demand for dissolved oxygen. The gas must be dissolved in the growth medium so that it can interact with the membrane-bound electron transport system to effect the oxidation of reduced pyridine nucleotide cofactors (Fig. 5.2). The fundamental problem in supplying sufficient oxygen is that oxygen is very insoluble in aqueous systems (Table 5.1). It should be noted from Table 5.1 that the solubility of oxygen decreases as the temperature increases. The solubility also decreases as the concentration of dissolved solutes increases. However, the solubility of oxygen in the growth medium can be increased by an increase in the oxygen pressure of the gas phase. A second problem is that in many fermentations, particularly those involving growth of filamentous organisms such as fungi or actinomycetes, the culture broth is extremely viscous and the controlling factor is transporting dissolved oxygen from the bulk liquid to the cell and not from the gas phase to the liquid phase.

For small-scale culture, e.g. less than 1 l, it is possible to supply adequate amounts of oxygen by growing the microbe in an Erlenmeyer flask which is agitated constantly. Agitation facilitates the transfer of oxy-

Table 5.1 The influence of temperature on the solubility of oxygen in pure water.

Temperature (°C)	Oxygen solubility in water (mg/l) at 1 atmosphere (101 kPa)
10	10.93
15	9.90
20	8.87
25	8.10
30	7.46
35	6.99
40	6.59

gen from the gas phase in the flask to the liquid phase. Provided that the volume of growth medium does not exceed 10−20% of the flask volume it is possible to achieve cell densities of 1−2 g dry wt/l before oxygen becomes limiting. Once the cell density increases beyond this level, oxygen utilization exceeds oxygen transfer to the liquid phase.

For large-scale culture and/or high cell densities the oxygen demand of the culture can be met only by forced aeration. In practice this is achieved by blowing air through the culture. The efficiency with which oxygen is transferred from air bubbles to the liquid phase principally depends on two functions: the surface area to volume ratio of the air bubbles and the residence time of the bubbles in the liquid. The smaller the bubbles, the greater the surface area to volume ratio and the greater the oxygen transfer. Similarly, the longer the bubbles remain in the liquid the greater will be the amount of oxygen which will diffuse from the bubble into the liquid. One way of decreasing bubble size is to introduce the air through a sparger with multiple small orifices rather than through a single large-bore orifice. A second way is to agitate the culture broth vigorously with a stirrer. There are two

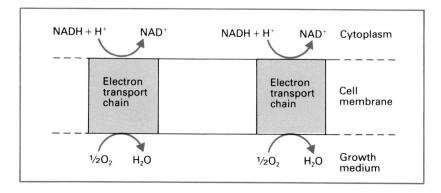

Fig. 5.2 The need for dissolved oxygen. Only oxygen in solution can interact with the cell membrane-bound electron transport chain.

additional advantages of agitation. The residence time of the bubbles is increased because they now follow a tortuous path rather than a straight path to the liquid surface. In addition, agitation facilitates oxygen transfer from the bulk liquid to the cell surface. Agitation is usually effected with a Rushton-type impeller (Fig. 5.3) since this is more efficient at bubble size reduction than a conventional marine propeller. The number of blades on the impeller and their dimensions relative to the fermenter dimensions greatly affect the efficiency of oxygen transfer.

BASIC FERMENTER DESIGN

From the above it should be apparent that the basic fermenter consists of a closed vessel fitted with an air inlet and an agitator. However, many other features are required and some of these are outlined below.

Baffles. Addition of baffles to the vessel walls can improve the efficiency of oxygen transfer by increasing the turbulence of the agitated culture medium.

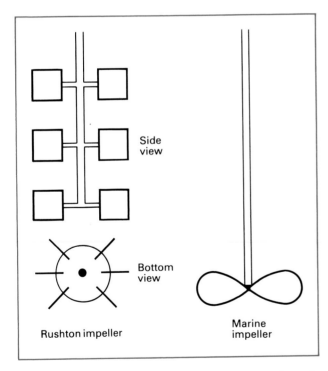

Fig. 5.3 Diagrammatic representation of a marine impeller and the six-bladed Rushton impeller.

Antifoam control. Agitation and aeration of the culture medium can result in excessive foaming, particularly at high cell densities and/or if complex growth supplements such as yeast or soya extract are present. Foaming can be detected by means of a simple probe placed above the culture medium. When the foam touches the probe it completes an electrical circuit and this activates a pump connected to a supply of an antifoaming agent.

pH control. The metabolism of most microorganisms results in a change in the pH of the culture medium with a concomitant change in the physiology of the microbe. It is usual practice to monitor continuously the pH of the fermentation broth with a pH probe *in situ* and to maintain a fixed pH by the addition of acid or alkali.

Temperature control. Microbes produce heat as they metabolize substrates. Exchange coils inside the vessel and/or a vessel jacket are used to counter this metabolic heat. As the fermentation proceeds the cell density increases, as does the heat output, and cooling water is circulated through the coils or jacket. Cooling water is used in an identical way to reduce the temperature of the fresh medium after it has been sterilized.

Addition ports. Provision must be made for the addition of the culture inoculum and media requirements.

CURRENT DESIGN

The above requirements are embodied in the basic stirred tank reactor design of fermenter as shown in Fig. 5.4. The material of fabrication is stainless steel and this must be of a high grade if it is not to corrode or leak toxic metal salts into the growth medium. The design of the fermenter shown in Fig. 5.4 is little changed from that developed in the 1940s for penicillin production, although the engineering is now to a far higher standard. The biggest change has been made to the agitators. Originally the agitator shaft was inserted through the roof of the vessel and the seal between the shaft and the vessel was a simple stuffing box; current practice is to use mechanical seals (see below) and bottom-driven agitators (Fig. 5.5). Since bottom-driven agitator shafts are shorter and are fully immersed in liquid there is much less 'whiplash' of the shaft and consequently fewer mechanical problems.

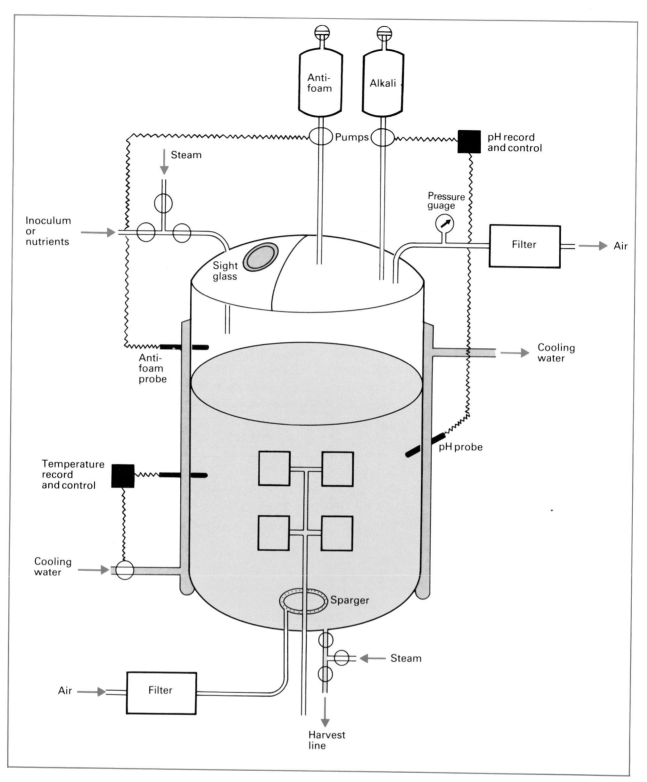

Fig. 5.4 Schematic representation of a stirred tank reactor. For clarity no seal is shown between the agitator shaft and the fermenter body and baffles have been omitted.

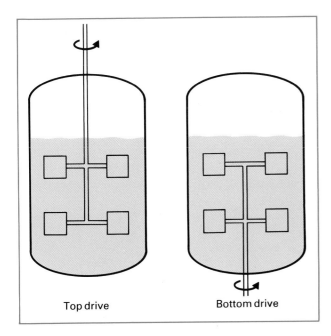

Fig. 5.5 Top- and bottom-driven agitator assemblies for fermenters. Note that the shaft length is shorter with a bottom-drive system.

ASEPTIC OPERATION

A key requirement for a successful fermentation is aseptic operation: that is, no contamination of the culture should occur. Thus the entire fermenter and ancillary equipment, as well as the growth medium, must be sterile before inoculation. In addition, the air supplied during the fermentation must be sterile and there must be no mechanical breaks in the fermenter that will allow the ingress of microorganisms.

All the equipment is cleaned with hot water before use and then sterilized *in situ* with steam. For sterilization to be successful the steam must reach all parts of the fermenter assembly, i.e. there must be no pockets of air. This requires judicious equipment design, particularly in the location of valves. In addition, all surfaces must be smooth and therefore welds, etc. must be polished; if any rough surfaces are present, however small, they can act as reservoirs of contaminating microorganisms. The growth medium can be heat-sterilized in the fermenter itself by passing steam through the cooling coils and jacket. Alternatively, the medium can be prepared in a separate vessel and passed through a continuous sterilizer, in essence a series of stacked hot plates, *en route* to the fermenter. Similarly, additives such as antifoaming agents must

be sterilized. The concentrated acids or alkalis used for pH control do not need to be sterilized.

The air supplied to the culture medium is sterilized by filtration. Many different kinds of filter material have been developed but in most instances the microbes in the air are not removed by size exclusion. Rather, the air follows a tortuous path through the filter matrix and the microbes are entrapped. Although bacteria and fungi are removed by filtration, bacteriophages in the air may not be and their presence in the fermenter can result in wholescale loss of the culture. Careful choice of filter design and air flow rate can minimize the problem as can the choice of air compressors. Some compressors generate sufficient heat to inactivate any bacteriophages present. It does help if a fermentation plant designed for large-scale culture of *Escherichia coli*, for instance, is not located downwind from a sewage treatment plant or a farmyard: both are rich sources of coliphages! Air leaving the fermenter may also need to be sterilized; this will certainly be necessary if recombinant organisms are being grown. This is not only a safety requirement but it prevents key microbial strains from being released into the environment and hence becoming freely available to competitors. Again, filtration is used.

The point at which the agitator drive shaft enters the vessel is a potential source of contamination. In the earliest fermenters stuffing boxes (Fig. 5.6) were used to minimize microbial entry. As the name implies, these were boxes stuffed with glass wool and phenolic grease. More recently, double mechanical seals (Fig. 5.6) have been introduced and the moving parts are lubricated with condensed steam to maintain sterility.

OPERATION OF FERMENTERS [1, 2]

The starting-point for any fermentation is a clean vessel which is sterilized and charged with sterile medium. The fermenter then has to be inoculated and the size of the inoculum is generally of the order of 1–10% of the total volume of medium; if it is any smaller, there may be a prolonged lag period before growth commences and the fermentation period will be unduly prolonged. An Erlenmeyer flask culture can be used as the inoculum for a small fermenter with a capacity of 10–20 l. Once the fermenter volume increases beyond this it is necessary to prepare the inoculum in a smaller fermenter. For a very large production fermenter a fermenter train (Fig. 5.7) may be needed to provide the inoculum. This greatly

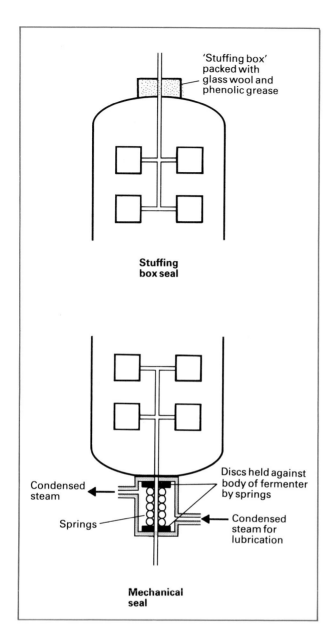

Stuffing box seal

'Stuffing box' packed with glass wool and phenolic grease

Mechanical seal

Condensed steam

Springs

Discs held against body of fermenter by springs

Condensed steam for lubrication

Fig. 5.6 The original 'stuffing box' seal and a modern mechanical seal used to prevent ingress of microorganisms where the agitator shaft passes through the body of the fermenter. Mechanical seals can be used with either top- or bottom-driven agitators. Stuffing boxes cannot be used with a bottom-drive system for the vessel contents would leak out.

increases the capital requirements of the process.

There are two major problems associated with the scale-up of fermentations. The first of these is culture stability. As outlined in Chapter 1, the classical method of strain development is repeated mutagenesis and culture selection. This eventually yields an enfeebled strain which overproduces the desired metabolite. When such high-yielding strains are cultivated they can be overgrown by more rapidly growing revertants which give reduced product yields. The chances of this happening increase as the scale of the fermentation increases. Ideally synthesis of the product should be 'switched off' until the final production fermenter is reached. With classical overproducing strains this is seldom possible. There is much less of a problem with genetically engineered strains because enzyme or protein synthesis can be controlled by judicious choice of promoter (see p. 26).

A second problem associated with scale-up is that many of the physical and chemical parameters which influence the behaviour of a microorganism are changed as the scale of operation is changed (Table 5.2; see also Box 5.1, p. 68). Thus operating conditions which are optimal for product formation at 100 l, for example, are not necessarily optimal at a much larger scale. Unfortunately it is too expensive and time-consuming to do a large number of fermentations at the production scale in order to find the ideal conditions. A number of techniques are used to overcome this problem, albeit not very satisfactorily, but they will not be discussed within the scope of this chapter. Such scale-up problems are not restricted to organisms that overproduce which are developed by classical mutation and selection procedures. Although many genetically engineered organisms are immune to these problems, again because of the use of controllable promoters, filamentous organisms are not. This is because organisms such as actinomycetes and filamentous fungi are shear-sensitive and this, in turn, affects their morphological filament size, the viscosity of the culture medium and the oxygen transfer coefficient.

CONTROL SYSTEMS

Many commercial aerobic fermentation processes are difficult to optimize and are not particularly reproduc-

Table 5.2 Variation in physical parameters on scale-up.

Scale-up criterion (parameter held constant)	Change on 1000-fold scale-up		
	Power	Power/ unit vol.	Impeller speed
Equal power/unit volume	10^3	1	0.22
Equal impeller speed	10^5	10^2	1.0
Equal impeller tip speed	10^2	0.1	0.1

Frozen stock → 10 ml culture → 200 ml culture → 3 l culture → 30 l culture → 300 l culture → (etc.)

Fig. 5.7 A fermenter 'train' used to provide a large inoculum for a production vessel.

ible. One cause of these problems is the use of complex nutrients such as starch, vegetable oil, soya meal, molasses and corn steep liquor. These nutrients are not chemically defined and there is substantial batch to batch variation, which can significantly affect product yield. There can even be substantial variation in contaminating substances in the bulk inorganic salts, e.g. phosphate, nitrates, etc., used in media preparation. These problems are compounded, for the reasons given in the previous section, if filamentous organisms are used. Ideally the fermentation technologist would like continuously to monitor the intracellular activities of the cell and to build up a profile of a standard fermentation. Then, if any deviation from the standard profile is detected in subsequent fermentations, corrective action may be taken by altering the environmental conditions.

If such *real time* process control is to be developed, there are two prerequisites. The first is suitable computer systems, and the appropriate hardware and software is now available. The second is the ability to measure intracellular activities on-line. To date the only such activity which can be measured on-line is the $NAD^+/NADH$ ratio and this is done fluorimetrically. Although there are assays for the ATP concentration inside the cell, e.g. the luciferase assay, these have to be done off-line. On-line measurements which are readily available are temperature, pH, dissolved oxygen and exhaust gas composition. The pH value and dissolved oxygen content of the medium provide essential information for alarm purposes, i.e. any deviation must be corrected, but would give a misleading picture of the process if used for control purposes. This is particularly true for the dissolved oxygen reading since this is a function of both medium viscosity and chemical composition and these change radically as the fermentation progresses.

Exhaust gas analysis can be done on-line and if a mass spectrometer is used much valuable physiological information can be obtained, e.g. respiratory quotient and the oxygen uptake rate. Thus, if an undesirable change in the oxygen uptake rate is detected, it is possible to decide which of the following strategies would have the least undesirable impact on the fermentation: alteration in the nutrient feed rate, the agitator speed, the air supply rate or the gas pressure inside the vessel.

BATCH VERSUS CONTINUOUS OPERATION

In batch processing the fermenter is filled with medium, sterilized and inoculated and the fermentation allowed to proceed. During the time course of the fermentation nutrients, acid or alkali, antifoam and air are supplied and product gases are removed. On completion, the fermenter is emptied and cleaned and prepared for a fresh run. In some instances this turnaround time between fermentations can be prolonged.

Box 5·1 Definition and significance of some terms used by biochemical engineers

Aspect ratio. The ratio of height to diameter of a fermenter. For a conventional stirred tank the ratio is between 1 : 1 and 4 : 1.

K_La This is the oxygen transfer coefficient. The rate at which oxygen is transferred from a gas bubble to the liquid phase can be expressed as:

$$\text{Rate} = K_La\,(c^* - c_L)$$

where a is the surface area of the bubble, c^* the concentration of oxygen in the bubble (equivalent to the partial pressure of oxygen), c_L is the dissolved oxygen concentration, and K_L a constant of proportionality. Since it is not possible to measure a with any reliability, it is usual to measure K_La. Note that if the value of c^* is increased by raising the air pressure inside the fermenter, the rate of oxygen transfer to the liquid will increase.

Liquid hold-up. The volume increase that occurs when gas is bubbled through a liquid.

Metre-cubed. Slang way of referring to cubic metres of fluid. For small volumes it is convenient to measure volume in litres but for larger volumes measurement is in cubic metres. One cubic metre is equivalent to 1000 l. Other useful figures are: 1 imperial gallon = 4.55 l; 1 US gallon = 0.83 imperial gallons = 3.79 l.

Metric ton. There are 1000 kg (equivalent to 2205 lb) in a tonne or metric ton whereas there are 2240 lb in an imperial or US ton.

Newtonian fluid. Shear forces are generated when two contiguous liquid or solid surfaces slide in opposite directions. A Newtonian fluid is one whose rate of shear is proportional to shear stress. The constant of proportionality is the fluid viscosity. Most fermentation media exhibit Newtonian fluid behaviour prior to inoculation and even when there is substantial growth of bacteria and yeast. Culture fluids containing filamentous microorganisms exhibit non-Newtonian fluid behaviour because they are much more viscous than the cell-free culture fluid. This complicates calculations of how much power needs to be imparted to the agitator.

Oxygen uptake rate. The rate at which cells utilize dissolved oxygen.

Power per unit volume. The power used to agitate a given volume of liquid. If the same power per unit volume is used to agitate two different sizes of vessel, it can be shown that the tip speed (q.v.) of the impeller will be greater in the larger vessel. This, in turn, will increase the shear forces on the microorganisms present.

Tip speed. The speed at which the edge of an impeller moves through a liquid (or space). Numerically it can be expressed as:

$$\text{tip velocity} = \text{rpm}\,(2\pi r)$$

where r is the distance of the impeller tip from the centre of rotation. Thus at any set rpm the greater the diameter of the impeller, the faster the tip moves through the liquid and the greater the shear force exerted on the liquid and any microorganisms it contains. Thus damage to mycelia increases as fermenters increase in size even if the rpm of the agitator is held constant.

Volumetric air flow rate. The volume of air supplied to a culture in a given time relative to the volume of that culture. Thus, at a flow rate of 1 VVM, one volume of air is supplied to an equal volume of culture fluid in one minute. If the volume of liquid in the fermenter were 100 l, then 100 l of air would be supplied in one minute.

Working volume. The maximum volume of a fermenter to which culture fluid can be filled before aeration and agitation commence. Space must be left for the volume expansion (liquid hold-up) which will occur on aeration and for any foam which will form.

In continuous fermentation raw materials are supplied to the vessel at a rate volumetrically equal to that at which product and spent medium are withdrawn (Fig. 5.8; see also Box 5.2). One advantage of continuous culture is that provided long runs, e.g. of many weeks' duration, can be achieved without contamination then there is no significant turnaround time. In this way the productivity of a particular vessel is increased. The extent of this increase will depend on the ratio of fermentation time to turnaround time: the smaller the ratio, the more advantageous is continuous operation. A second advantage is that the growth rate of the cells can be maintained at a value that is optimal for product formation.

Continuous culture is ideally suited for the production of biomass or simple metabolites, such as ethanol, whose synthesis is proportional to cell density. It is not suitable for the production of other metabolites such as amino acids and antibiotics, whose synthesis is not associated with growth. In addition, the strains used in these processes are too unstable for continuous operation (see p. 66). The most widespread use of continuous culture is in the effluent disposal industry (see p. 97), where waste organic materials (substrate) are converted into microbial cells (biomass).

NOVEL REACTOR DESIGNS

The stirred tank reactor (Fig. 5.4) is the workhorse of the fermentation industry because of its reliability and versatility. Nevertheless some novel designs have been introduced in recent years. Considerable

---

Box 5·2 Theory and practice of continuous culture

If unicellular microbes are growing exponentially in a nutrient medium then the rate of change of cell number (n) with time (t) will be proportional to the number of cells. That is:

$$\frac{dn}{dt} = \mu n$$

where μ is the constant of proportionality, otherwise known as the growth rate constant. Eventually some nutrient will become limiting, the growth rate will decrease and finally all growth will cease. In continuous cultures fresh nutrients are constantly added to the medium and an equal volume of culture removed, i.e. the total culture volume remains the same. Nutrient addition can be controlled in two ways. In *turbidostats* the turbidity of the culture is monitored constantly and when it rises above a pre-set value fresh medium is added. In *chemostats* growth is controlled by limiting the supply of a critical growth factor, i.e. some nutrient is limiting. If fresh medium is added to a chemostat at a rate of F l/h and the chemostat has a volume of medium V l, then the rate of loss of organisms from the fermenter is given by:

$$-\frac{dn}{dt} = \frac{F}{V} n = Dn.$$

The ratio F/V is known as the dilution rate (D)

and has the dimension of reciprocal hours (h^{-1}). At the steady state, dn/dt should be zero, in which case $\mu = D$, i.e. the rate of growth is identical to the dilution rate, which is a function of the rate at which fresh growth medium is added to the vessel. Clearly, if D is allowed to exceed μ_{max}, cells will be lost at a rate faster than they can grow and *washout* will occur. A useful approximation is that when the dilution rate is $0.1\,h^{-1}$ the cells have a doubling time of 7 h and when the dilution rate is $0.7\,h^{-1}$ the cells have a doubling time of 1 h.

In chemostat culture all nutrients except one are present in excess and the level of the limiting nutrient determines the cell density. The nutrients that can be used to limit growth include the sources of carbon, nitrogen, phosphorus, sulphur or metal ions. Different limiting nutrients affect the cell physiology in different ways.

The advantage of turbidostat and chemostat culture over batch culture is that the cells are in a constant environment. However, cells in turbidostat culture grow at the maximum rate permissible with the selected medium; cells in chemostat culture are always starving for some nutrient and are not growing at the maximum rate possible except when the value of D approaches that of μ_{max}.

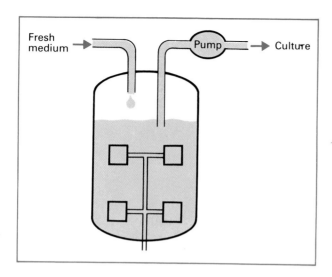

Fig. 5.8 Addition of fresh medium and removal of culture fluid in a continuously operated fermenter in the chemostat mode. The arrangement shown is just one of many used to ensure a uniform volume of culture fluid.

power is needed to drive the agitators used on large fermenters and to compress the air supplied. This leads to high utilities costs, principally for electricity. One way of reducing the power requirements is to use the air supplied to agitate the culture. In an *airlift* fermenter (Fig. 5.9) the air is introduced at high velocity at the bottom of the vessel. The use of a draught tube enables the liquid to circulate with considerable turbulence. For efficient oxygen transfer it is necessary for the fermenter to have a height to diameter ratio (10:1−40:1) that is much greater than with a conventional stirred tank reactor (1:1−4:1).

As a gas bubble ascends a column of liquid the hydrostatic pressure on it decreases and the volume of the bubble increases. Since there is now a decrease in the surface area to volume ratio there is a decrease in the rate of oxygen transfer from the gas phase to the bulk liquid. Furthermore, in a well-agitated fermenter over 50% of the oxygen supplied is utilized. Thus, as the bubble ascends the column the partial pressure of oxygen within it will decrease and this too will decrease the rate of oxygen transfer to the liquid. A novel reactor design which overcomes these disadvantages is the *deep-shaft reactor* (Fig. 5.9). Here the air is introduced at high velocity at the top of the reactor and forces liquid downwards. Although oxygen is utilized during the downward passage the bubble size will diminish because of the hydrostatic pres-

sure and should give rise to a more uniform oxygen transfer rate.

Both the airlift fermenter and the deep-shaft reactor can be operated continuously. With both of them the entire vessel contents are essentially homogeneous, both spatially and temporally. By contrast, in a bubble column reactor (Fig. 5.9), which is a variant of the airlift fermenter, the contents of the vessel are not homogeneous. Since there is no mixing, the medium moves through the fermenter as a 'plug' and its composition varies spatially. By dividing the fermenter into sections it is possible to vary the physicochemical environment at will, e.g. pH 7.0 and 30°C in one section and pH 5.0 and 40°C in another section.

DOWNSTREAM PROCESSING [3, 4]

Once a fermentation is complete, it is necessary to recover the desired end-product. At a minimum this will involve separation of the cells from the fermentation broth but may also include purification of metabolites with or without cell disruption. Such operations are referred to as *downstream processing*. Where the desired product is heat-labile it is necessary to cool the fermentation broth prior to cell removal. This can be achieved either by passing ice-cold water through the fermenter coils and/or jacket or by passing the fermenter contents through a heat exchanger.

Some cells rapidly settle out of suspension once aeration and agitation of the fermenter broth cease and this settling may be assisted by the addition of flocculating agents. Many of these flocculating agents are polyelectrolytes, which neutralize the charge on the microbial cell surface, which keeps cells in suspension by electrostatic repulsion. Where cell settling occurs, or can be induced to occur, then cell removal is relatively inexpensive for the fermentation broth can be decanted or filtered off. Flocculation is an essential property of yeasts used in the production of alcoholic beverages. Usually yeast sedimentation occurs towards the end of the fermentation but in traditional ale production in the UK the yeast cells float to the surface and are skimmed off.

Where cell settling does not occur, cell removal can be effected by centrifugation. Efficient cell removal is favoured by large particle sizes, a large density difference between the particle and the fluid, a low fluid viscosity, a large centrifuge radius and a high angular velocity. The particle size and density and the viscosity of the fermentation liquor are usually outside

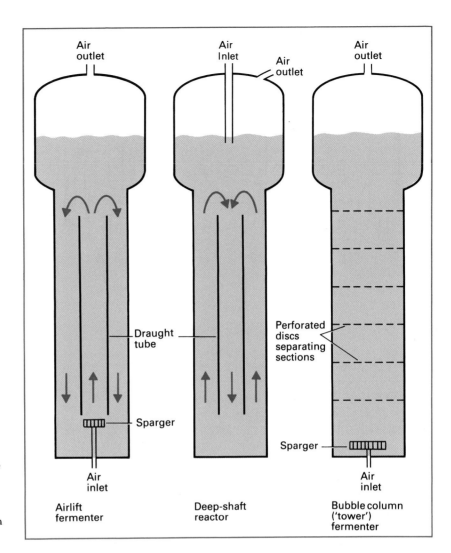

Fig. 5.9 Some novel reactor designs (see text for details). Of the three designs illustrated, only the airlift fermenter is used in both the batch and continuous modes; the other two are always used in the continuous mode.

the control of the fermentation technologist. Additionally, the radius of the centrifuge bowl cannot be increased indefinitely since the mechanical stress on the bowl increases with the square of the radius and safe limits are quickly reached. Thus for large volumes of fluid, centrifuges which can operate in a continuous-flow mode are required. Many different designs are available but only those that can operate at high angular velocities (Fig. 5.10) are of use for removing cells or cell debris. When operating continuously these centrifuges are more efficient at particle removal when the liquid to be clarified is flowing through them slowly. This is because the time that each particle is exposed to the centrifugal force is increased.

An alternative to centrifugation is *ultrafiltration*. The term ultrafiltration describes processes in which particles significantly greater in size than the solvent are retained when the solution is forced through a membrane of very fine pore size, usually less than 0.5 µm. Not surprisingly with such fine pore sizes there is tremendous resistance to the passage of liquid and acceptable flow rates are achieved only by the application of pressure and by the use of filter designs that have a high surface area to volume ratio. Extracellular proteins may or may not be retained by the membrane depending on the membrane pore size.

The clarified fermentation liquor will contain microbial metabolites and extracellular enzymes, and many methods are available for their recovery, including ion exchange chromatography, precipitation and solvent extraction. The use of ion exchange resins is particularly desirable since they can be used for concentration as well as having high capacity and being cheap and reusable.

Fig. 5.10 Three commonly used designs of continuous-flow centrifuge.

If the desired product is retained within the cell, then cell disruption will be necessary to effect its release. Although many methods of cell disruption are used in the laboratory most of them cannot be used on a large scale. Two which are suitable for large-scale use are high-pressure homogenization and the use of high-speed bead mills. In high-pressure homogenization cell suspensions move from a high-pressure environment to a low-pressure environment by passage through a fine orifice. The cells rupture by explosive decompression and the process is facilitated by their impaction at high velocity on components of the homogenizer (Fig. 5.11(a)). Cell disruption, as measured by release of protein from the cells, is increased with increasing temperature and pressure and is independent of cell concentration. The desirability of high-pressure units is best illustrated by the fact that the efficiency of disruption is approximately proportional to the cube of the pressure.

High-speed bead mills are agitators consisting of a central shaft on which a set of circular discs is eccentrically mounted to form a helical array (Fig. 5.11(b)). The tank is filled with very small (0.45 mm diam.) glass beads prior to charging with the cell slurry. It is the abrasion caused by the rapid motion of these beads which ruptures the cells. The efficiency of disruption increases with increasing agitator speed and bead loading and decreasing bead size, temperature and cell concentration.

After cell disruption the cell debris needs to be removed. As with removal of intact cells, the most widely used methods are centrifugation and ultrafiltration.

Fig. 5.11 Schematic representations of equipment used for continuous disruption of cells (see text for details). (a) High-pressure homogenizer. (b) High-speed bead mill (beads not shown for clarity).

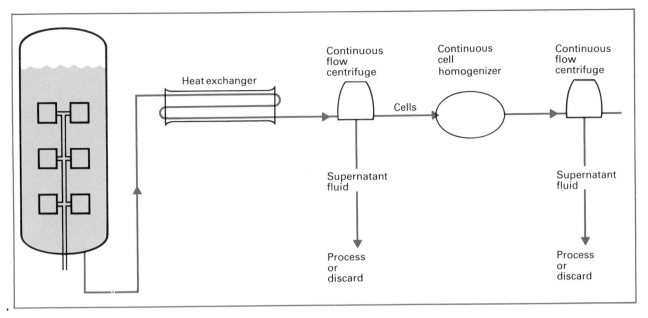

Fig. 5.12 Schematic representation of typical operations in the processing of a fermentation broth. If the desired product is in the broth, the cells are discarded. If the product is inside the cells, the cells may need to be homogenized prior to product extraction.

SAFETY CONSIDERATIONS [5]

There have been few safety concerns associated with the more traditional large-scale commercial fermentations. This is because most of the organisms used are non-pathogenic. In recent years, however, there has been considerable debate about the potential hazards associated with genetically engineered organisms. These hazards are purely conjectural and the evidence to date suggests that recombinant organisms are no more dangerous than their traditional counterparts. Nevertheless, it is prudent for the fermentation technologist to take heed of public concern when designing production facilities. In actual fact there is little hazard associated with the fermentation itself for a vessel designed to keep contaminating organisms out also serves effectively to retain potential pathogens! Design features such as double mechanical seals on the agitator shaft and filters on the exhaust gas outlet are particularly useful. There are a number of specific modifications which can be made to decrease any risk; for example, careful design of sampling ports can eliminate aerosol formation and provision can be made to contain any spills which occur. Also, all effluent from the fermenter, including that produced during cleaning, can be passed through a continuous sterilizer or to a kill tank.

Potentially there are more hazards associated with downstream processing than with the fermentation. Figure 5.12 shows some typical unit operations associated with a fermentation. Heat exchangers pose few problems provided care is taken to prevent freezing of the culture fluid; this could result in fracture of the exchanger following thermal expansion. The centrifuges used are likely to be of the continuous-flow type and these are notorious for generating aerosols in the immediate environment. Hermetically sealed models are available and should be used and all vents on them should be fitted with microbiological filters. Finally, all the equipment should be connected with high-quality piping, including high-quality welding, and it should be possible to sterilize all of it.

Further reading

GENERAL REFERENCES

Pirt S.J. (1975) *Principles of Microbe and Cell Cultivation*. Blackwell Scientific Publications, Oxford.
Wang D., Cooney C.L., Demain A.L., Dunnill P., Humphrey A.E. & Lilley M.D. (1979) *Fermentation and Enzyme Technology*. John Wiley & Sons, New York.

CITED REFERENCES

1 Buckland B.C. (1984) The translation of scale in fermentation processes: the impact of computer process control. *Bio/technology* **2**, 875–83.

2 Buckland B.C., Brix T., Fastert H., Gbewonyo K., Hunt G. & Jain D. (1985) Fermentation exhaust gas analysis using mass spectrometry. *Bio/technology* **3**, 982–8.

3 Tutunjian R.S. (1985) Scale-up considerations for membrane processes. *Bio/technology* **3**, 615–26.

4 Van Brunt J. (1986) Fermentation economics. *Bio/technology* **4**, 395–401.

5 Bennett A.M., Hill S.E., Benbough J.E. & Hambleton P. (1991) Monitoring safety in process biotechnology. In *Genetic Manipulation: Techniques and Applications* (Ed. by J.M. Grange, A. Fox & N.L. Morgan), pp. 361–376. Blackwell Scientific Publications, Oxford.

6 The Commercial Exploitation of Microorganisms

Microorganisms have been subjugated by man in many different ways. These encompass the production of industrially important materials including fine chemicals (e.g. pharmaceuticals) and bulk chemicals, the manufacture of single-cell protein from diverse substrates, the processing or recycling of waste materials, and coping with what is loosely called the energy crisis. One convenient way of subdividing the different microbial processes is shown in Table 6.1 and each of the different categories will be discussed in turn. Many of these processes are not new. On the contrary, they have been established for two, three or even more decades. Currently many of them are undergoing rapid development and this is a direct result of the introduction of recombinant DNA technology. Other processes are being introduced now for the first time for their existence depends solely on the ability to manipulate genes *in vitro*. One could say that the industry is in a 'ferment'!

Production of whole cells (biomass)

There are a number of instances in which microbial cells themselves are the desired end-product and these are listed in Table 6.2. Most of them have rather specialized applications, e.g. the production of *Penicillium roquefortii* spores. The characteristic flavour of

Table 6.1 The different biotechnological applications of microorganisms.

1 Production of whole cells (biomass)
2 Production of low-molecular-weight compounds
  (a) production of primary metabolites
  (b) production of secondary metabolites
  (c) transformation of organic compounds by non-growing cells
3 Production of high-molecular-weight compounds
  (a) polysaccharides
  (b) lipids
  (c) proteins
4 Processes dependent on general microbial metabolism
  (a) degradation/oxidation of effluents and noxious wastes
  (b) mineral extraction

Table 6.2 Some applications of microbial cells.

Organism	Application
Bacillus thuringiensis and related organisms	Microbial insecticide
Lactobacillus sp., *Streptococcus cremoris* and related species	Starter cultures for the manufacture of dairy products, e.g. yoghurt, cheese
Penicillium roquefortii and related species	Inocula for the production of blue-veined cheeses
Rhizobium sp.	Inoculants for adding to legume seeds to promote nodulation and nitrogen fixation
Pseudomonas syringae	Creation of artificial snow. Ice-nucleation-defective mutants for the prevention of frost damage to crops
Many different organisms	Single-cell protein production

blue cheeses is due to the metabolic products of *P. roquefortii* growing in the milk fats. Strain selection has resulted in the identification of several isolates with particularly desirable properties for cheese manufacture. These include the ability to produce adequate amounts of low-molecular-weight ketones and alcohols, which are responsible for the characteristic flavour, together with good proteolytic activity, which is essential for texture. Production of the spores is limited to a few, small companies but has potential for expansion because cultures grown on milk fats can be used as sources of blue cheese flavouring; naturally produced flavourings are superior to mixtures of purified chemicals because of the flavour subtleties of trace amounts of other metabolites.

INOCULANTS [1]

The *Rhizobium* inoculants are of far greater economic value than the blue cheese inocula described above but their use is largely confined to Australia and New Zealand. The reason for this is twofold. First, forage legumes such as clover and lucerne (alfalfa) were not indigenous to these countries but were introduced by the early settlers. Since they were not indigenous crops there was no reservoir in the soil of the *Rhizobium* sp. whose infection of root hairs leads to the beneficial nitrogen-fixing symbiosis which makes legumes so agriculturally important. One way of overcoming this is to add the appropriate rhizobia to the soil along with the seeds. This is in fact what is done. However, in climates such as that of Australia, the added rhizobia tend not to survive from year to year and hence inoculation is required in each planting season.

Despite the economic value of inoculation there are probably fewer than 10 companies worldwide, all of them small operations, producing these *Rhizobium* inoculants. Although considerable progress has been made in recent years in understanding the molecular basis for the legume—microbe symbiosis, it is unlikely that gene manipulation will have a significant impact on the development of improved strains for inoculation. Better strains will come about by selection of improved isolates already existing in nature. Although there has been much talk about the development of rhizobia that infect other plants, such as cereals, it is more likely that in the long term the biochemical apparatus for nitrogen fixation will be introduced into plants by manipulation of the plant genome (see Chapter 12).

An organism whose commercial exploitation has begun only recently is *Pseudomonas syringae*. A surface component of this bacterium catalyses the formation of ice nuclei and is finding use in the formation of 'artifical' snow on ski slopes. The presence of this organism on the surface of leaves can promote frost damage to crops, so considerable benefit might be obtained if plants were sprayed with ice-nucleation-

defective mutants. Currently this subject is surrounded by controversy regarding the environmental implications and is discussed in more detail in Chapter 14.

INSECTICIDES [2, 3, 4, 5]

The widespread use of agrochemicals has allowed man to achieve unprecedented control over pests and has contributed towards today's high agricultural productivity. This is reflected in the fact that Europe for the first time is a food exporter and has built up extensive reserves of grain and other produce. However, the potential toxic effects on man and the environment and the rapid development of resistance in the target pests has led to a reconsideration of the real value of chemical insecticides. Consequently interest is turning to certain microorganisms which are natural pesticides (Fig. 6.1). Although over 100 bacteria, fungi and viruses that infect insects have been described, only a very few are in commercial production (Table 6.3). The most widely used is *Bacillus thuringiensis*, which produces intracellularly a proteinaceous crystal toxic to insects. After ingestion by susceptible insects the crystalline toxin dissolves in the alkaline environment of the midgut, where it inhibits ion transport leading to cessation of feeding and death. Much still remains to be learned about the mechanism of toxicity; for example the nature and potency of the toxin crystals varies amongst bacterial serotypes and different strains are required to control different pests. Once more is known about the biological basis of toxicity it may be possible to use protein engineering to develop a single strain which is pathogenic to a wide variety of insects. A step in this direction has been taken already. Two

Fig. 6.1 Death of an aphid due to infection with the fungus *Verticillium lecanni*. (Courtesy of Microbial Resources Ltd.)

Table 6.3 Examples of some microbial insecticides in commercial production.

Organism	Used to control
Bacillus thuringiensis	
Kurstak strain	Caterpillars
H-14 strain	Whitefly and mosquitoes
Bacillus popillae	Japanese beetles
Bacillus penetrans	Nematodes (eelworms)
Hirsutella thompsonii	Citrus mite in Florida
Verticillium lecanni	Glasshouse aphids in UK
Metarhizium anisopliae	Spittle-bug in Brazilian sugar cane and pastures
Beauvaria bassiana	Colorado beetle in USSR

crystal protein genes were fused and expression of the gene fusion in *E. coli* produced a biologically active protein with a toxicity spectrum that overlapped those of both contributing crystal proteins. Recombinant DNA technology has also been used to transfer the genetic information for toxin production to a *Pseudomonas* sp. which is commonly found associated with the roots of maize plants. The objective here is to control root-boring pests in the US corn belt. The same genes have been transferred to plants and this is described in Chapter 13.

STARTER CULTURES [6, 7, 8]

A major application of microbial cells is the addition of starter cultures to milk for the production of fermented products such as cheese, yoghurt, sour cream, etc. The basic reason for converting milk into products such as cheese is to preserve the valuable nutrients in milk. Thus cheese is made because this affords a means of converting an easily perishable food (milk) into a stabler product (cheese). This is brought about in a variety of ways: a decrease in water content and water activity, a decrease in pH caused principally by the formation of lactic acid, the depletion of food components which could be used by spoilage and pathogenic microorganisms and the production of 'natural' antimicrobial substances such as D-leucine. Another reason for making fermented milk products is to provide both textural and flavour variety in foods.

The use of lactic starter cultures for the manufacture of dairy products is a relatively new business. For thousands of years man depended upon the natural contaminating flora in milk for the synthesis of lactic acid and flavours and the formation of the desired product. However, the widespread thermal processing

of milk, e.g. pasteurization, to destroy pathogens such as the causative agents of tuberculosis and brucellosis alters the microbial flora of milk and it no longer naturally undergoes a lactic fermentation. Consequently starter cultures have to be added and these are generally frozen or lyophilized cell concentrates of either pure cultures of lactic acid bacteria or selected mixtures. When these bacteria are added to milk they cleave the lactose ('milk sugar') to glucose and galactose and these are metabolized to lactic acid (Fig. 6.2). The acid coagulates the milk proteins, principally casein, to form a continuous solid curd in which fat globules,

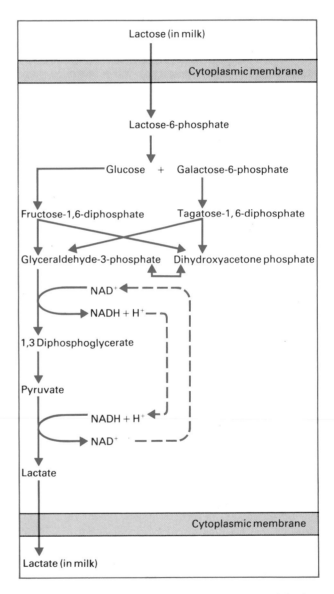

Fig. 6.2 The metabolic interconversions of lactose which lead to lactic acid formation. Not all the intermediates are shown.

water and water-soluble materials are entrapped. The curd is then pressed to expel the fluid whey, shaped and left to mature. During maturation subtle flavours are formed by enzymic conversion of the curd contents and by microbial metabolism.

Growth of lactic acid bacteria in milk is dependent not only on phosphogalactosidase activity but also on protease activity. These bacteria are nutritionally fastidious and require an exogenous supply of amino acids and peptides. In order to grow in milk they must degrade casein and the extent of this proteolysis contributes to the flavour of the final product.

Gene manipulation has much to offer the dairy microbiologist. For example, a major problem in the production of fermented milk products is slow acid formation, which can result in off-flavours, extended process times or complete loss of production. Slow acid formation can be caused by inhibition of the bacteria by antibiotic residues in the milk, loss of the ability to grow on lactose or to degrade casein, or infection with bacteriophages. Now that the genetics of the lactic acid bacteria are receiving much attention the construction of antibiotic-resistant derivatives should not be too difficult. Both the ability to grow on lactose and to degrade casein are plasmid-mediated traits and increasing plasmid stability is feasible. Alternatively, it should be possible to clone the relevant genes and force their integration into the chromosome. Finally, if the number of copies of the plasmid were increased, this could result in more rapid acid formation and hence decreased process times.

Attack by bacteriophages is probably the biggest source of problems for the dairy industry. Bacteriophages abound in milk and the dairy environment generally and their complete elimination is not possible. A number of plasmids which encode resistance to bacteriophages have been identified in lactic streptococci. Resistance can result from inhibition of phage adsorption, restriction of phage DNA and inhibition of phage maturation. In most instances resistance has been determined for only a few phage isolates. Nevertheless it is hoped that by means of gene manipulation a cassette of genes can be constructed which will embody all the above resistance mechanisms and greatly reduce the susceptibility of lactic streptococci to a wide range of bacteriophages. Of course, the use of such strains for the preparation of food would require regulatory approval.

Just as starter cultures are required for the production of fermented milk products, so too are they required for the preparation of fermented meat products, such as salami. The initial bacterial flora of meat supplied for sausage production is more variable today because of modern methods of transporting, pretreatment and slaughter of animals. The starter cultures used are mixtures of lactobacilli and streptococci, to produce lactic acid which acts as a preservative, and micrococci. The nitrate- and nitrite-reducing activities of the micrococci convert myoglobin to nitrosomyoglobin; the latter is essential for red colour formation.

SINGLE-CELL PROTEIN [9]

The term 'single-cell protein' (SCP) refers to cells, or protein extracts, of microorganisms grown in large quantities for use as human (*food*) or animal (*feed*) protein supplements. Although SCP has a high protein content, it also contains fats, carbohydrates, nucleic acids, vitamins and minerals. Its use in food presents a problem because humans have a limited capacity to digest nucleic acids and additional processing is necessary. The animal feed market is more attractive because not only is less processing of the product require but the regulatory process is less stringent.

Over the years a large variety of raw materials has been used for SCP production including carbon dioxide, methane, methanol, ethanol, sugars, petroleum hydrocarbons and industrial and agricultural wastes. These feedstocks have been used with a variety of different microorganisms including algae, actinomycetes and other bacteria, yeasts, filamentous fungi and higher fungi. Commercial success depends, among other things, on the ready availability of cheap substrate and its efficient conversion into cellular protein. It is worth noting that one of the economic realities of biotechnology is that waste products, which by definition are readily available, are cheap only so long as nobody wants them.

For economic, political and technical reasons SCP has not become the important source of food protein that was forecast originally. The developed countries do not need SCP for they have a plentiful supply of high-quality protein as a result of the agricultural improvements of the last forty years. Evidence of this is provided by the butter and grain 'mountains' which exist in the EEC and USA. SCP would be a useful food additive in tropical countries, where traditional food products are high in carbohydrate and low in protein. This chronic shortage of protein leads to physical and mental deterioration. In countries of the Middle East, and in Africa south of the Sahara, the land is too arid to produce sufficient food of any type regardless of its protein content. Thus in both the arid and tropical regions the production of SCP would be advantageous but the countries concerned cannot afford the initial capital investment. They also lack the technical expertise and support facilities to maintain production once established. One country which might favour SCP production is the USSR since it has the unique combination of technical expertise coupled with poor agricultural productivity and a lack of hard currency.

Even as a feed supplement SCP has not been a success because of cost considerations. Feedstocks, particularly those associated with oil production, have generally increased in price, whereas those from competitive sources of protein, such as soybean meal or fishmeal, have decreased in price. Even if oil prices drop, they can just as easily rise again because of world politics. Fluctuating commodity prices do not encourage large capital investment.

SCP production facilities have been constructed in most of the developed countries but many of these are no longer operating for the reasons stated above. The most impressive is the one built by Imperial Chemical Industries in the north of England. In this plant, which has the world's largest continuous operating fermenter, the bacterium *Methylophilus methylotrophus* was grown on methanol and ammonia. The product, which is called Pruteen, contains 80% crude protein and has a high vitamin content. Although twice as nutritious as soybean meal it is not economic to produce.

Recombinant DNA technology has been used in an interesting way to try and improve the efficiency of SCP production. Ammonia must be assimilated for cellular growth and there are two different routes whereby this occurs (Fig. 6.3). *M. methylotrophus* does not possess the enzyme glutamate dehydrogenase and uses instead the ATP-dependent GS/GOGAT pathway. This represents a potential source of methanol wastage. When the glutamate dehydrogenase gene from *E. coli* was cloned and expressed in *M. methylotrophus* there was a 4–7% increase in the efficiency of carbon conversion.

BACTERIAL VACCINES [10, 11, 12, 13]

A clinical use of bacterial cells is for the production of vaccines. Some vaccines, e.g. those for diphtheria

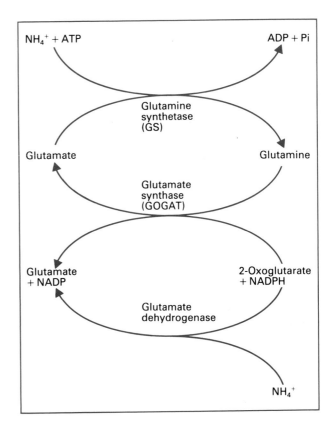

Fig. 6.3 The two routes of ammonia assimilation which are found in microorganisms.

and tetanus, are purified exotoxins which have been rendered non-toxic by treatment with heat or formalin. It has been suggested that impurities incorporated during formaldehyde treatment are partly responsible for the adverse effects sometimes seen with hyperimmunization with tetanus toxoid. When the toxin is treated with the protease papain a fragment from the C-terminus (fragment C) is produced. Fragment C derived in this way has been shown to be non-toxic and capable of immunizing mice and guinea pigs. However, preparation of fragment C is hazardous since it involves the purification of large quantities of tetanus toxin followed by carefully controlled papain digestion. Fortunately fragment C can now be prepared in *E. coli* and is just as efficacious. However, most bacterial vaccines, such as those for cholera, whooping cough, plague, typhoid, paratyphoid fever and typhus fever, are simply suspensions of killed bacteria. After growth the bacteria are killed with heat or chemical agents such as phenol, acetone or formalin. Only the vaccine for immunization against tuberculosis uses live cells, in this case an attenuated strain of *Myco-bacterium bovis*. Whereas vaccines prepared from live, attenuated organisms often give lifelong protection, vaccines prepared from killed cells have to be injected repeatedly according to carefully balanced immuniz-ation schedules. Recently a novel way of attenuating *Vibrio cholera* has been developed. Using recombinant DNA technology a portion of the gene for cholera enterotoxin was deleted. The modified protein pro-duced is immunogenic but non-toxic and has been used as a single-dose oral vaccine in humans.

An alternative approach to the development of live vaccines is to start with the food-poisoning organism *Salmonella typhimurium*. This organism can be attenu-ated by the introduction of lesions in the *aro* genes, which encode enzymes involved in the biosynthesis of aromatic amino acids, *p*-aminobenzoic acid and enterochelins. Whereas doses of 10^4 wild-type *S. typhimurium* reproducibly kill mice, *aro* mutants do not kill mice when fed orally, even when doses as high as 10^{10} organisms are used. However, the mutant strains can establish self-limiting infections in the mice and can be detected in low numbers in organs such as the liver and spleen. Such attenuated strains of *S. typhimurium* are particularly attractive as carriers of heterologous antigens because they can be delivered orally and because they can stimulate humoral, se-cretory and cellular immune responses in the host.

Already a wide range of heterologous antigens have been expressed in such vaccine strains.

Production of low-molecular-weight compounds

Microorganisms are used to produce a wide variety of low-molecular-weight compounds and some examples are given in Table 6.4. These compounds can be sub-divided (Fig. 6.4) into those whose production is as-sociated with growth (*primary metabolites*) and those whose synthesis occurs after growth ceases (*secondary metabolites*). Primary metabolites can be further sub-divided into those such as vitamins and amino acids, which are normally produced in quantities sufficient only for cell growth, and those such as ethanol and lactic acid, which are produced in large quantities because they are normal metabolic end-products. Microbes are also used to effect chemical transform-ations; that is, the desired product is not a normal metabolite of the cell but is produced as a result of enzymatic conversion of an unusual substrate added to the culture medium. Often such substrates cannot support growth; they simply undergo *bioconversion*.

METABOLIC END-PRODUCTS [14]

Metabolic end-products such as ethanol, acetone, butanol and lactic acid are the traditional products of

Compound	Use	Producing organism
Antibiotics		
Colistin	Antibacterial agent	*Bacillus colistinus*
Griseofulvin	Antifungal agent	*Penicillium griseofulvum*
Avermectin	Antihelminthic agent	*Streptomyces avermitilis*
Pharmacologically active agents		
Ergot alkaloids	Vasoconstriction	*Claviceps purpurea*
Cyclosporin	Immunosuppression	*Tolypocladium inflatum*
Mevinolin	Cholesterol reduction	*Aspergillus terreus*
Vitamins		
Cyanocobalamin (B_{12})	Pernicious anaemia	*Propionibacterium shermanii*
Amino acids		
Glutamic acid	Flavour enhancer	*Corynebacterium glutamicum*
Steroids		
11-α-hydroxylation	Production of therapeutically useful steroids	*Rhizopus nigricans*
Other		
Sorbose	Vitamin C production	*Gluconobacter suboxydans*
Indigo	Textile dye	*Escherichia coli*

Table 6.4 Examples of the diversity of commercially useful low-molecular-weight substances produced by microorganisms.

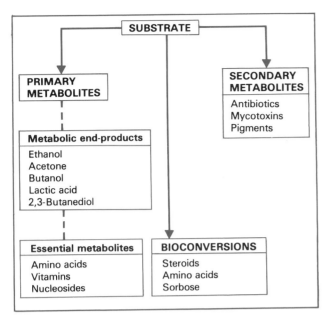

Fig. 6.4 The different classes of low-molecular-weight compounds synthesized by microorganisms.

the fermentation industry. They are also fermentations in the strictest sense because they are formed as a result of the anaerobic metabolism of sugar substrates. In the absence of oxygen as terminal electron acceptor different cells regenerate NAD^+ from NADH by different means (Fig. 6.5), some of which lead to the formation of commercially important chemicals. Ethanol is different in that it has two functions. Like the others it is a commodity chemical but, more importantly from an economic viewpoint, its production is also the basis of an enormously varied alcoholic beverage industry (Table 6.5). Over 16 billion imperial gallons (20 billion US gallons) of beer alone are produced annually, making it the single most important fermentation product.

As with SCP, the fermentative production of these metabolic end-products is subject to economic and political pressures. Thus their production declined after the Second World War with the growth of the petrochemical industry, which was based on cheap oil supplies. As the price of oil has risen there has been a

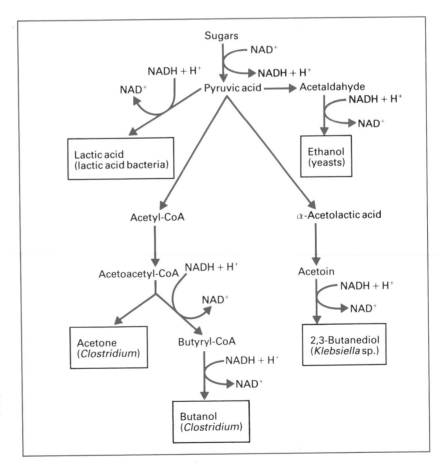

Fig. 6.5 The formation of commercially useful metabolic end-products. Note that pyridine nucleotide cofactors are reduced during the conversion of sugars to pyruvate and subsequently oxidized by further metabolism of pyruvate.

Table 6.5 The origins of the different kinds of alcoholic beverages.

Alcoholic beverage	Origin
Non-distilled	
Beer	On germination, starch in barley grains is converted to sugar, which is extracted by boiling in water to produce wort and this is fermented
Cider	Fermentation of apple juice
Wine	Fermentation of grape juice
Sake	Starch in steamed rice is hydrolysed with *Aspergillus oryzae* and the sugars released are fermented with yeast
Distilled	
Whisky (Scotch)	Distillation of alcohol produced from barley
Whiskey—Irish	Pot still whiskey produced from alcohol derived from a mixture of barley, wheat and rye. Grain whiskey produced from alcohol derived from maize
—Rye	Produced from alcohol derived from rye
—Bourbon	Produced from alcohol derived from maize
Rum	Distillation of fermented molasses, a by-product of sugar cane refining
Vodka	Distillation of alcohol produced from any non-grain carbohydrate source, e.g. potatoes
Gin	Distillation of alcohol derived from maize or rye and redistillation in presence of herbs and juniper berries
Tequila	Distillation of fermented extracts of Mexican cactus

move back to fermentation as a means of production. Thus the percentage of industrial alcohol production which is based on fermentation has risen in the last 10 years from less than 10% of total world demand to more than 20%. This increase in fermentative production is not without its consequences. Much of it takes place in Brazil and is dependent on sugar cane for the provision of substrate. To provide sufficient sugar cane plantations large tracts of forested land have been cleared and this could be ecologically disastrous. Much of the pressure in Brazil to produce alcohol is politically driven and politics may also influence the fortunes of the acetone–butanol fermentation. The political embargo on the export of oil to South Africa may trigger redevelopment of the acetone–butanol fermentation there to provide the raw materials for its manufacturing industry.

One advantage of the solvent fermentations over SCP production is that a less sophisticated facility is required to produce them. This is certainly true when one compares the processes used for acetone–butanol fermentation (see p. 4) and brewing with those required to grow microorganisms aseptically on methane or methanol. Therefore it is not inconceivable that suitable fermentation plants could be established in the developing countries.

One metabolic end-product of *Saccharomyces cerevisiae* is carbon dioxide, whose formation by baking yeast is an essential step in leavening of dough. A novel baking yeast has been constructed in which the normal promoters for maltose permease and maltase, which are subject to glucose repression, are replaced by the constitutive promoter from the yeast alcohol dehydrogenase gene. The constitutive synthesis of the proteins involved in maltose transport and degradation eliminates the delay during dough leavening of maltose gene induction and consequent fermentation. Usually the presence of even a trace amount of glucose represses the maltose system and in many doughs the glucose content is very high.

ESSENTIAL PRIMARY METABOLITES

Essential primary metabolites are those compounds which microorganisms synthesize because they are essential for cell growth. A number of these compounds are of economic importance (Table 6.6). Microorgan-

Table 6.6 Economically important, essential primary metabolites produced by fermentation.

Metabolite	Use
Amino acids	
L-Glutamate	Flavour enhancer
L-Threonine	Feed supplement
L-Lysine	Feed supplement
L-Phenylalanine	Manufacture of Aspartame (artificial sweetener)
L-Tryptophan	Feed supplement
Vitamins	
Riboflavin (vitamin B_2)	Food supplement
Vitamin B_{12}	Food supplement and feed additive
Nucleotides	
Inosine-5′-monophosphate	Flavour enhancer
Guanosine-5′-monophosphate	Flavour enhancer
Pigments	
β-carotene	Precursor of vitamin A

isms do not naturally overproduce these compounds because if they did so they would be at a competitive disadvantage relative to strains producing only enough for growth. This is because overproduction wastes sources of carbon and energy. To obtain overproducing strains it is necessary to subvert the cellular regulatory mechanisms. Metabolic regulation is achieved in two ways: by *feedback inhibition* of the activity of key enzymes by the end-product of a biosynthetic pathway and *repression* of the synthesis of the enzymes in a pathway by the end-product. Overproduction can also be facilitated by eliminating competing biochemical reactions. These principles can best be explained by reference to a specific example, that of threonine over-production.

A diagrammatic representation of threonine bio-synthesis is shown in Fig. 6.6. Starting with a wildtype strain of *E. coli* that produces no excess threonine, a threonine-overproducer was obtained in three steps.

1 A mutant excreting 1.9 g/l was obtained by isolating a mutant resistant to the amino acid analogue α-amino, β-hydroxy valerate. In such mutants the enzyme homoserine dehydrogenase is no longer sensitive to feedback inhibition by threonine.

2 A mutant unable to convert threonine to isoleucine was selected following mutagenesis of the organism isolated in step 1 above. This mutant has two important properties. First, threonine is not converted to other compounds which reduce the overall yield. Second, the low internal concentration of isoleucine is insufficient to repress the synthesis of the enzymes converting homoserine to threonine and isoleucine. The threonine yield at this stage was 4.7 g/l.

3 The strain from step 2 was further mutated such that it now had a requirement for methionine. This raised the threonine yield to 6 g/l. Clearly, before such strains

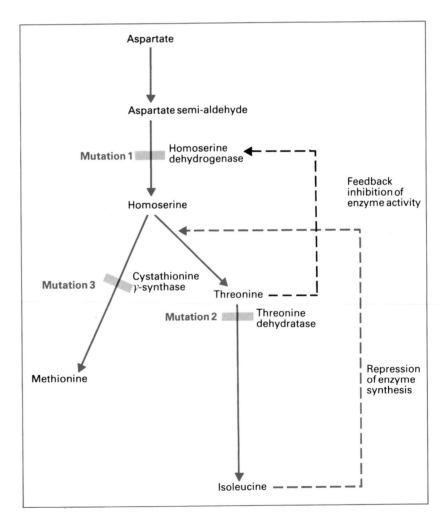

Fig. 6.6 Genetic manipulation of amino acid biosynthetic pathways leading to overproduction of threonine. Mutation 1 renders homoserine dehydrogenase insensitive to feedback inhibition by threonine (broken black arrow). Mutations 2 and 3 inactivate the enzymes indicated. See text for details.

in steps 2 and 3 will grow, small amounts of methionine and isoleucine need to be added to the culture medium.

Another way to circumvent the regulatory controls of the cell is to alter the permeability of the cytoplasmic membrane. This enables the desired product to leak from the cell, thereby reducing the intracellular concentration. In *Corynebacterium glutamicum*, which yields over 90 g glutamate/l, membrane permeability can be altered by biotin deficiency since biotin is essential for fatty acid synthesis and thus membrane integrity.

The removal of both the normal cellular regulatory controls and competing biochemical pathways can be achieved readily by conventional genetic techniques. Recombinant DNA technology also has a role to play; for example, one enzymic step may be rate-limiting, thus reducing the metabolic flux through the pathway. By simply cloning the relevant gene such that there are multiple copies inside the cell the enzyme level can be amplified. However, there is an even better use for gene manipulation: clone the genes for all the enzymes in a biosynthetic pathway such that their synthesis is under the control of a single, well-characterized promoter, e.g. the *lac* promoter (see p. 16). Not only is the normal cellular repression of enzyme synthesis eliminated, synthesis of the enzymes can be controlled by the fermentation technologist. Where the producing organism is cultured in a series of fermenters of increasing size (see Fig. 5.7), synthesis can be repressed until the final production vessel is reached. In this way there is less problem of overgrowth of energetically fitter, non-producing mutants.

SECONDARY METABOLITES [15, 16, 17, 18]

Secondary metabolites are molecules synthesized by microorganisms late in the growth cycle (Fig. 6.7). They are not required for growth and their real function is not known. Since many of them, e.g. antibiotics, are inhibitory to other organisms they may impart an ecological advantage on the producing organism. The best known of the secondary metabolites are the antibiotics. Others include mycotoxins, ergot alkaloids, the widely used immunosuppressant cyclosporin, and fumagillin, an inhibitor of angiogenesis and a suppressor of tumour growth. Over 2500 antibiotics have been described of which the majority are produced by actinomycetes. The diversity of molecular structures is impressive (Fig. 6.8) and many of them are produced

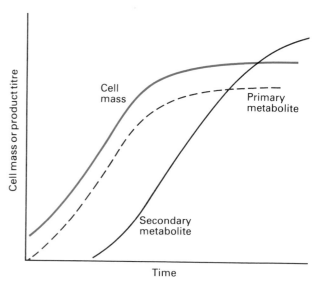

Fig. 6.7 Kinetics of formation of primary and secondary metabolites in relation to cell growth.

as mixtures of related compounds. Thus there are over 20 penicillins, 20 actinomycins, 10 polymyxins, 10 bacitracins, 3 neomycins, etc.

Unlike primary metabolites such as amino acids, secondary metabolites as products were developed at a time when little was known about their pathways of biosynthesis. Even today details are unclear for many of them. Consequently two approaches have been used to obtain enhanced yield of product. One of these is the random mutation and selection procedure described earlier (see p. 6) and an example is shown in Fig. 6.9. The other involves screening of hundreds of culture medium additives as possible precursors of the desired product. Occasionally a precursor is identified that increases production of the secondary metabolite; for example, α-aminoadipate addition stimulates penicillin biosynthesis. Alternatively, the precursor may direct the formation of one specific desirable product ('directed biosynthesis'). Thus phenylacetic acid addition promotes the formation of benzylpenicillin over other penicillins. Both approaches have been very successful but both are labour-intensive. Occasionally they can be combined. For example, where addition of an amino acid has been shown to be stimulatory then mutational efforts should be focused on removing regulatory controls on biosynthesis of that amino acid (as described in the previous section).

Development of superior antibiotic-producing strains has been hampered in two ways. First, the biosynthetic pathway is often not known, making it difficult to

Fig. 6.8 Structures of some representative antibiotics.

apply selective mutation techniques. Second, the regulatory controls on synthesis are not known, particularly that mechanism which delays synthesis until the end of the cellular growth phase. This is where recombinant DNA technology can be extremely helpful. The genes controlling entire biosynthetic pathways can be cloned. Once cloned, the DNA can be sequenced and this permits identification of the number of genes involved and is a good indication of the number of biosynthetic steps. Since the protein products of the cloned genes are overexpressed their purification is facilitated. By deleting one or more of the genes in the pathway, intermediates will accumulate and these can be identified by conventional analytical procedures,

e.g. mass spectrometry. In this way a picture of the biosynthetic pathway can be constructed. More important, the normal regulatory circuits can be eliminated and the cloned genes placed under the control of a known promoter, thus facilitating fermentation development.

One of the interesting findings resulting from the cloning of genes encoding antibiotic production is that synthesis of key enzymes is dependent on the presence of positively acting regulatory elements rather than on a release from repression by elements acting negatively (see Box 6.1). Positive regulatory genes are much more likely to be inactivated by mutagenesis, giving a non-producing phenotype, than to be made more effective in stimulating production. Thus it is

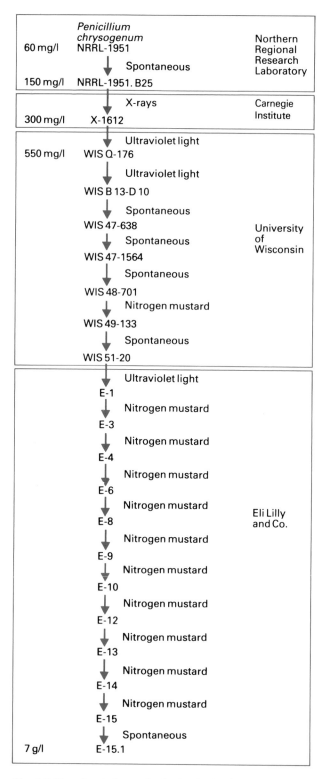

60 mg/l	*Penicillium chrysogenum* NRRL-1951	Northern Regional Research Laboratory
	↓ Spontaneous	
150 mg/l	NRRL-1951.B25	
	↓ X-rays	Carnegie Institute
300 mg/l	X-1612	
	↓ Ultraviolet light	
550 mg/l	WIS Q-176	
	↓ Ultraviolet light	
	WIS B 13-D 10	
	↓ Spontaneous	
	WIS 47-638	University of Wisconsin
	↓ Spontaneous	
	WIS 47-1564	
	↓ Spontaneous	
	WIS 48-701	
	↓ Nitrogen mustard	
	WIS 49-133	
	↓ Spontaneous	
	WIS 51-20	
	↓ Ultraviolet light	
	E-1	
	↓ Nitrogen mustard	
	E-3	
	↓ Nitrogen mustard	
	E-4	
	↓ Nitrogen mustard	
	E-6	
	↓ Nitrogen mustard	
	E-8	Eli Lilly and Co.
	↓ Nitrogen mustard	
	E-9	
	↓ Nitrogen mustard	
	E-10	
	↓ Nitrogen mustard	
	E-12	
	↓ Nitrogen mustard	
	E-13	
	↓ Nitrogen mustard	
	E-14	
	↓ Nitrogen mustard	
	E-15	
	↓ Spontaneous	
7 g/l	E-15.1	

Fig. 6.9 Use of mutation and selection in the development of improved penicillin-producing strains of *P. chrysogenum* by mutation and selection.

likely that the beneficial mutations obtained by random mutagenesis (Fig. 6.9) and selection simply improve the supply of metabolites and cofactors to the relevant pathway. Such beneficial changes would not readily have been obtained through the application of recombinant DNA techniques. On the other hand, cloning techniques are ideally suited to the control of positive regulatory genes and should nicely complement the conventional mutagenesis approach.

Just as novel proteins can be produced by recombinant DNA techniques (see Chapter 4), so too can novel antibiotics be produced. The *Streptomyces coelicolor* gene cluster encoding the biosynthesis of the isochromanequinone antibiotic actinorhodin has been cloned. When the cloned genes were introduced into a variety of other *Streptomyces* spp. producing different isochromanequinones at least three new antibiotics were detected. Clearly actinorhodin, or one of its precursors, is a novel metabolite in these other *Streptomyces* spp. and is subject to further or different enzymatic modifications. Yet another way of producing novel antibiotics is the provision of unusual substrates to enzymes involved in antibiotic biosynthesis. A key step in the synthesis of β-lactams (penicillin and cephalosporins) is the cyclization of a tripeptide precursor. When the purified enzyme is fed novel tripeptides, novel β-lactams are produced.

BIOCONVERSIONS [19, 20]

Bioconversions (or biotransformations) are processes in which microorganisms convert a compound to a structurally related product. They comprise only one or a small number of enzymatic reactions as opposed to the multi-reaction sequences of fermentations. Although hundreds of different bioconversions have been described, they are only used commercially when conventional chemical approaches are too costly or difficult—for example, when stereoselective transformation is required, when there are no functional groups near the atom to be modified, or when only one of many identical functional groups in a molecule is to be modified. Some examples are shown in Fig. 6.10. The ultimate in specificity are the steroid bioconversions (Fig. 6.11). Their utility can best be illustrated by the synthesis of cortisone. The original, entirely chemical synthetic route comprised 37 steps and the end-product cost $200/g. The introduction of a microbial bioconversion reduced the number of steps to 11 and the cost dropped to $6/g. Acetic acid bacteria

Regulation of transcription is necessary to avoid non-essential protein synthesis. The simplest type of gene regulation is that found in *negatively controlled* gene clusters such as the *E. coli lac* operon. In this case the operon consists of four elements: a repressor gene, operator and promoter sites, and the structural genes encoding the proteins whose synthesis is regulated. The repressor is synthesized constitutively and binds tightly to the operator, thereby restricting access of RNA polymerase to the promoter. In the presence of enzyme substrate (inducer), the repressor undergoes a conformational change such that it no longer recognizes the operator. Once the repressor is removed from the operator, RNA polymerase can bind to the promoter and mRNA can be synthesized.

In the case of biosynthetic operons the substrate of the first enzyme is not an inducer. Rather, the end-product of the biosynthetic pathway acts as a co-repressor. An example is the operon of *E. coli* responsible for synthesis of tryptophan. If tryptophan is present in the growth medium there is no need for the operon to be turned on. If tryptophan is absent, the operon must be fully functional. This regulation is accomplished by means of a repressor that is inactive unless tryptophan is bound to it. Regulation using the end-product as co-repressor is efficient when the role of the gene cluster is to synthesize a molecule, whereas the use of a substrate as inducer is of value when the gene cluster specifies a degradative pathway.

Many gene clusters are regulated by *positive control* elements instead of by repressors. Positive control elements can act either by binding directly to the promoters, to proteins which bind near promoters, or to RNA polymerase itself. Most positively controlled gene clusters are relatively complex but the principle of all of them is basically the same: the substrate of a key enzyme stimulates the interaction of RNA polymerase and the promoter.

are particularly useful for transformations of polyhydric alcohols, the most important being the conversion of sorbitol to sorbose (Fig. 6.10), an essential raw material for the chemical synthesis of vitamin C (Fig. 6.12).

Although bioconversions can be conducted with growing cells, many are carried out with resting or killed cells or even spores. The advantages of using non-growing cells are threefold. First, very high substrate loadings can be used, e.g. tens of grams of substrate per litre, and such elevated substrate levels usually inhibit growth. Second, if washed cells are used, there will be no contaminating substances present. These two features make product isolation very easy. So too does the fact that substrate conversion efficiencies of 85–100% are readily achieved. Finally, many enzymes have pH and/or temperature optima far removed from those of the intact cell. In bioconversions these can be optimized to increase the reaction rate.

Another feature of bioconversions is that they are easy to scale up since the only parameter of interest is the level of the enzyme mediating the transformation.

This makes bioconversions obvious candidates for the application of recombinant DNA technology. Not only can the enzyme levels be increased in existing processes but processes which were previously uneconomic can be reconsidered. Thus the levels of aspartate aminotransferase in wild-type *E. coli* are too low to make this organism of much use for L-amino acid synthesis from α-keto acids (Fig. 6.10). By cloning the relevant gene, enzyme levels can be raised to over 20% of total soluble protein and bioconversion becomes feasible. Once the gene is cloned it is possible to use site-directed mutagenesis (see p. 47) to produce enzyme variants with enhanced thermostability or altered substrate profile. It is also possible to develop novel synthetic routes and the best example is provided by the manufacture of vitamin C. The conventional process starts with glucose and comprises one microbiological and four chemical steps (Fig. 6.12(a)). By cloning in *Erwinia* a gene from *Corynebacterium* the process can be simplified to a single microbiological and a single chemical step (Fig. 6.12(b)). Another variant of this concept of new synthetic routes is provided

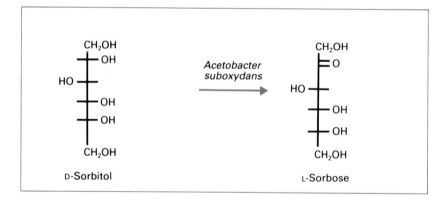

Fig. 6.10 Representative examples of commercial bioconversions.

by the microbial synthesis of indigo. Cloning of a single *Pseudomonas* gene, that encoding naphthalene dioxygenase, resulted in the generation of an *E. coli* strain able to synthesize indigo from tryptophan (Fig. 6.13).

Production of high-molecular-weight compounds

Microorganisms are used to produce two groups of high-molecular-weight compounds (macromolecules): polysaccharides and proteins. The commercial production

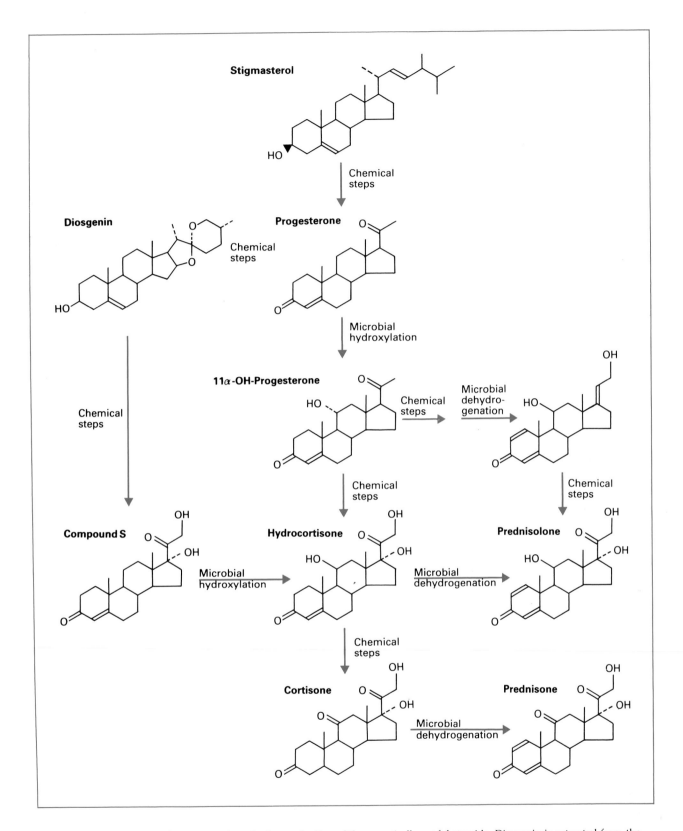

Fig. 6.11 The involvement of microorganisms in the production of therapeutically useful steroids. Diosgenin is extracted from the roots of the Mexican barbasco plant. Stigmasterol is extracted from soybeans.

Fig. 6.12 Simplified route to vitamin C (ascorbic acid) developed by cloning in *Erwinia* the *Corynebacterium* gene for 2,5-diketogluconic acid reductase. (a) Classical route to vitamin C. (b) The simplified route to 2-ketogulonic acid, the immediate precursor of vitamin C.

of both has increased dramatically in the last 5–10 years but for different reasons. Recombinant DNA technology has had an impact on protein production because now it is possible to overproduce proteins which are normally synthesized in trace amounts and consequently have been too costly and difficult to isolate. By contrast, polysaccharides have traditionally been isolated from plants, particularly seaweeds. This has disadvantages including high labour costs for harvesting and extraction, limited availability of the sources, a supply that can be affected adversely by climate, and seasonal variation in polysaccharide composition. Microorganisms can provide a constant and reliable supply of a polymer with relatively uniform chemical and physical properties.

MICROBIAL POLYSACCHARIDES

The industrial value of polysaccharides lies in their capacity for altering the rheological properties of aqueous solutions, either through gelling or by the alteration of their flow characteristics. Many polysaccharides are thixotropic, i.e. the solutions are characterized by high viscosity at low stress and decreased viscosity when stress is applied. Such characteristics make them particularly useful in a whole variety of

industries (Table 6.7). Pullulan has potential applications in the cosmetic and food industries but its greatest use may be as a biodegradable plastic. The clear films and plastic containers currently in use are made from petroleum-based products, which are not biodegradable.

The greatest single potential market for polysaccharides is the oil industry. Conventional oil extraction technologies can recover only about 50% of the world's subterranean oil reserves; the balance is either trapped in rock or too viscous to pump. The application of microorganisms, or their products, to aid in the recovery of this oil is known as *microbial-enhanced oil recovery* (MEOR). For example, biological compounds that could be injected into wells include surfactants and viscosity decreasers to release trapped oil, and viscosity enhancers to push oil out of crevices. Xanthan gum can be used as a viscosity enhancer and emulsan as a surfactant. Like many biologically produced polymers, emulsan exhibits a specificity that is not generally observed in chemically synthesized materials; the emulsifying activity of emulsan is substrate-specific, acting only on hydrocarbons that have both aliphatic and cyclic components.

Polysaccharides are found outside the microbial cell wall and membrane, hence the use of the term *exopoly-*

Fig. 6.13 The production of indigo by *E. coli* carrying a cloned naphthalene dioxygenase gene. Tryptophan in the medium is converted to indole by endogenous tryptophanase.

saccharide. Their synthesis is favoured by an excess of carbon substrate in the growth medium, particularly if nitrogen is limiting. Although they are mostly produced in batch culture, chemostat culture could be used. In this instance, as with biomass production, genetic instability is not a problem. Furthermore, one can vary the composition of the polysaccharide by varying the limiting nutrient; for example, the polysaccharide may be composed mostly of neutral sugars

if nitrogen is limiting but when metal ions are limiting there can be a high proportion of sugar acids (e.g. glucuronic acid).

With many microorganisms there are probably over 100 enzymatic steps directly or indirectly involved in polysaccharide biosynthesis. This is because microbes have the ability to produce several polysaccharides of different sugar composition and sequence. Additionally, subtle chemical alterations to one or more sugar residues can have dramatic effects on the physicochemical properties of the molecule. Thus for the introduction of process and product improvements it is likely that classical genetic techniques will initially be more useful than recombinant DNA technology.

POLYHYDROXYALKANOATES [31]

Many bacteria synthesize intracellular reserve materials known as polyhydroxyalkanoates (PHAs) of which poly(3-hydroxybutyrate) (PHB) is the best known. It has been known for over 20 years that PHB is a biodegradable thermoplastic material. More recently, microbial PHAs have been discovered whose properties are superior to PHB. The first PHA consumer product was launched in 1990—a biodegradable shampoo bottle constructed from a copolymer of 3-hydroxybutyrate and 3-hydroxyvalerate. Currently such a copolymer is expensive to produce but costs might be reduced now that the relevant genes have been cloned.

PRODUCTION OF PROTEINS [21, 22, 23, 24, 25, 26, 27, 28, 29, 30]

The microbial production of proteins is exploited commercially in two ways: the production of bulk enzymes and the production of much smaller quantities of proteins of therapeutic value. Bulk enzymes have a variety of applications (Table 6.8). Their manufacture is a well-established business and by the end of the 1980s the market was about 100 000 tonnes with a value of $1 billion. Fewer than 20 enzymes comprise the larger part of this market and only a proportion of them are produced by microorganisms. Because enzymes are direct gene products, theoretically they are good candidates for improved production through recombinant DNA technology. However, many of the production processes are already high-yielding, e.g. in excess of 20 g/l, and significant increases may not be possible. The use of gene manipulation

Polysaccharide	Producing organism	Uses
Xanthan gum	*Xanthomonas campestris*	1 Food additive for stabilizing liquid suspensions and gelling soft foods, e.g. ice cream, cheese spreads 2 Lubrication in, for example, toothpaste preparations 3 Enhanced oil recovery
Gellan	*Pseudomonas* sp.	1 Solidification of food products
Emulsan	*Acinetobacter calcoaceticus* *Arthrobacter*	1 Cleaning oil spills 2 Enhanced oil recovery
Pullulan	*Aureobasidium pullulans*	1 Biodegradable material for food coating and packaging
Dextrans	*Leuconostoc mesenteroides*	1 Blood expander 2 Adsorbents for pharmaceutical preparations

Table 6.7 Commercially available microbial polysaccharides and their uses.

Enzyme	Source	Applications
α-amylase	*Aspergillus oryzae* *Bacillus amyloliquefaciens* *Bacillus licheniformis*	Preparation of glucose syrups Removal of starch sizes Liquefaction of brewing adjuncts
β-glucanase	*Aspergillus niger* *Bacillus amyloliquefaciens*	Liquefaction of brewing adjuncts Improvement of malt for brewing
Glucoamylase	*Aspergillus niger* *Rhizopus* sp.	Starch hydrolysis
Glucose isomerase	*Arthrobacter* sp. *Bacillus* sp.	High-fructose corn syrup
Lactase	*Kluyveromyces* sp.	Removal of lactose from whey
Lipase	*Candida lipolytica*	Flavour development in cheese
Pectinase	*Aspergillus* sp.	Clarification of wines and fruit juices
Penicillin acylase	*Escherichia coli*	Preparation of 6-aminopenicillanic acid
Protease, acid	*Aspergillus* sp.	Calf rennet substitute
Protease, alkaline	*Aspergillus oryzae* *Bacillus* sp.	Detergent additive Dehairing of hides
Protease, neutral	*Bacillus amyloliquefaciens* *Bacillus thermoproteolyticus*	Liquefaction of brewing adjuncts
Pullulanase	*Klebsiella aerogenes*	Starch hydrolysis

Table 6.8 Sources and applications of some microbial enzymes.

techniques to modify proteins, as discussed in Chapter 4, may be more important because they could yield enzymes with improved properties. For example, a lowering of the pH optimum of glucose isomerase used in the production of high-fructose corn syrups would reduce the product discoloration ('browning reaction' or 'caramelization') that occurs with the normal enzyme. There is also the potential for increased thermostability. Cloning could also be used to provide alternative sources of enzymes which are in short supply. One such enzyme is rennet (chymosin), which is obtained from the stomach of young calves. This enzyme is used widely in the cheese industry to facilitate curdling of the milk (see p. 78). Currently demand

outstrips supply and a variety of microbial proteases ('microbial rennet') are used but these are of inferior quality since they generate unwanted flavours. The gene for chymosin has been cloned and commercial use is dependent only on regulatory approval.

One protein which is not an enzyme but which will be produced in bulk is bovine growth hormone. Administration of this hormone to cows significantly increases their milk yield. The hormone also has an effect on fish. Weekly injection of the hormone into young salmon results in a doubling in the rate of weight gain but it remains to be seen if this practice will be adopted by fish farms.

Perhaps the most glamorous aspect of recombinant

DNA technology is its use to construct microorganisms which synthesize human proteins with therapeutic potential. Many of these proteins are synthesized in trace amounts in the body, particularly if they are hormones, and before the advent of gene manipulation many had not been purified or even isolated in crude form. A list of those currently being investigated is given in Table 6.9. So far only a limited number have been approved for sale although many more are undergoing clinical trials. Initially, applications are made to

Table 6.9 Some human proteins with therapeutic potential.

Protein	Application	Current status
Human growth hormone (somatotropin)	Pituitary dwarfism	Approved for sale
Human insulin	Diabetes	Approved for sale
Interferon-α2b	Hairy cell leukaemia, genital warts and other applications	Approved for sale
Interferon-α2a	Hairy cell leukaemia, Kaposi's sarcoma and other applications	Approved for sale
Erythropoietin	Anaemia associated with kidney dialysis and AZT treatment of AIDS	Approved for sale
Tissue plasminogen activator	Myocardial infarction	Approved for sale
Hepatitis B coat protein	Vaccination	Approved for sale
Granulocyte colony stimulating factor	Neutropenia arising from cancer chemotherapy	Approved for sale
Interleukin-2	Cancer therapy	Approved for sale
Consensus interferon	Cancer therapy	Late clinical trials
Interferon-γ	Rheumatoid arthritis and cancer therapy	Late clinical trials
Interferon-β	AIDS therapy	Late clinical trials
Superoxide dismutase	Free radical damage of reperfusion, renal transplants	Late clinical trials
Factor VIII	Haemophilia	Late clinical trials
Lung surfactant protein	Respiratory distress syndrome	Late clinical trials
Tumour necrosis factor	Cancer therapy	Clinical trials
Epidermal growth factor	Healing of ulcers	Clinical trials
Fibroblast growth factor	Healing of ulcers	Clinical trials
Relaxin	Facilitation of childbirth	Early clinical trials

the regulatory authorities for a single disease state and, once approved, permission is sought for additional clinical indications. Thus interferon-α was initially approved for hairy cell leukaemia but is now used in the treatment of 16 different cancers. Despite this it is not a blockbuster drug: at the end of 1989 world-wide sales were less than $300 million. Already, too, some recombinant proteins are being abused. For example, erythropoietin is attractive to athletes because it stimulates red blood cells to increase their output of haemoglobin.

Many of the molecules listed in Table 6.9 are secreted in their natural host as glycoproteins (see Box 6.2). When synthesized in microbes they either will not be glycosylated or will exhibit a different glycosylation pattern. Thus the recombinant-derived molecules will be analogues. When used clinically the problems encountered in their administration will be those of any analogue: different spectrum of bioactivity, different *in vivo* distribution, different circulatory lifetime and different immunogenicity from the native form and, perhaps, new properties. Methods have been developed for covalent attachment of mono-methoxy polyethylene glycol to proteins and such modified molecules have an increased half-life in animal models. Nevertheless, they are still analogues. Other kinds of analogues are the second-generation molecules produced by protein engineering, which were described in Chapter 4.

Attempts have also been made to use genetic engineering to facilitate virus vaccine production. Thus the genes for the major subunits of a number of viruses, e.g. foot and mouth disease virus, have been cloned in *E. coli* but the results have been disappointing: the peptides thus produced have been poorly immunogenic. One notable exception is a hepatitis B vaccine prepared in yeast and which is on sale in a number of countries. Approaches using recombinant animal cells are more promising and are discussed on p. 124.

Regardless of whether a bulk enzyme or a therapeutic protein is being produced, downstream processing will involve protein purification. Standard biochemical techniques are used, e.g. ion exchange chromatography, and protein engineering can be used to facilitate purification, as the example of polyarginine-tailing shows (see p. 55). However, there is a fundamental difference in the purity requirements of the two classes of protein. With bulk proteins the major criterion is activity: the greater the activity of the protein the more useful it will be. Purity is much less important and if some of the enzyme is denatured during processing it is of little consequence provided the specific activity does not fall below a minimally acceptable level. By contrast, most of the therapeutically useful proteins will be administered parenterally and they must be of the highest quality (Table 6.10.)

When foreign proteins are synthesized in microorganisms the ends of the molecules can be 'nibbled' by proteases and such partially degraded molecules must be removed for they could become immunogenic if therapy were prolonged. An added complication is that, in *E. coli* at least, many foreign proteins are deposited in inclusion bodies (Fig. 6.14). The protein has to be extracted from these inclusions using denaturing agents and the protein subsequently refolded. This is a process akin to recreating native egg white from a meringue! Recently it has been shown that, for some proteins at least, their correct folding and assembly is facilitated by a class of proteins known as molecular chaperones. The archetypal chaperonin is the GroEL protein of *E. coli*, which facilitates the folding of certain proteins without becoming part of their final structure. If significant protein loss is not to occur during downstream processing, care must be taken to exclude microorganisms at all stages of purification. Finally, the overall process has to be conducted according to the code of *good manufacturing practice* (GMP) (see Box 6.3).

A newer use of cloned proteins comes from work on the cloning of human adrenergic receptors in *E. coli*. Such cloned receptors retain the binding properties of the corresponding receptors in mammalian tissues but are easier to use in pharmacological studies. Undoubtedly they will find use in the pharmaceutical industry for the rapid screening of synthetic molecules as potential drugs. Already a recombinant human interleukin-1 (IL-1) receptor antagonist has been

Table 6.10 Specification for therapeutic proteins to be administered parenterally.

Greater than 95% purity
Microheterogeneity below specified level
Contaminating DNA less than 10 pg/dose
Endotoxin below specified level
Toxic chemicals (used in purification) below specified level
Specific activity above a minimum level and preferably maximum possible
Absence of microorganisms

Box 6·2 The glycosylation of proteins

The secreted and cell-surface-associated polypeptides of higher organisms are subject to a variety of post-translational modifications involving carbohydrate moieties (Diagram 1). The presence of oligosaccharides not only influences the physical properties of a protein but also its physiological properties, e.g. bioactivity, biodistribution, immunogenicity and circulatory lifetime.

Diagram 1 N-glycosylation involves the covalent attachment of an oligosaccharide moiety to the NH_2 groups of asparagine in Asn-X-Ser/Thr triplets via an N-glycosidic bond. N-linked oligosaccharides are generally located at β-turns and most glycoproteins carry two or more N-linked oligosaccharides. O-glycosylation involves the covalent attachment of an oligosaccharide moiety to hydroxyl groups of serine (or threonine) of the polypeptide via an O-glycosidic bond. In the covalent attachment of a polypeptide via a glycosylphosphatidylinositol (GPI) membrane anchor, the C-terminal amino acid residue is attached to a glycan moiety through an ethanolamine phosphate bridge; the glycan moiety is attached to diacylglycerol through an inositol phosphate bridge. An increasing number of cell surface glycoproteins have been found to be anchored into the lipid bilayer via such a group. Diagram reproduced courtesy of Professor R. Dwek and the editors of *Trends in Biotechnology*.

developed which reduces the lethality of endotoxin-induced shock. IL-1 has been implicated as an important mediator of septic shock because it can induce tachycardia and hypotension and act synergistically with tumour necrosis factor to cause tissue damage and death.

Box 6·3 The code of good manufacturing practice (GMP)

GMP guidelines can and should be applied in any manufacturing operation. As far as biotechnology is concerned these guidelines are most appropriate to the production of therapeutic proteins. According to the UK guide to GMP, 'there should be a comprehensive system, so designed, documented, implemented and controlled, and so furnished with personnel, equipment and other resources as to provide assurance that products will be consistently of a quality appropriate to their intended use'. The basic requirements of GMP are that:

1 all manufacturing processes should be clearly defined, and known to be capable of achieving the desired ends;

2 all necessary facilities should be provided, including:

    (a) appropriately trained personnel;
    (b) adequate premises and space;
    (c) suitable equipment and services;
    (d) correct materials, containers and labels;
    (e) approved procedures (including cleaning procedures);
    (f) suitable storage and transport;

3 procedures are to be written in instructional form, in clear and unambiguous language, and are to apply specifically to the facilities provided;

4 operators should be trained to carry out the procedures correctly;

5 records are to be kept during manufacture (including packaging) which demonstrate that all the steps required by the defined procedures are in fact taken and that the quantity and quality produced are those expected;

6 records of manufacture and distribution which enable the complete history of a batch to be traced should be retained in a legible and accessible form;

7 a system must be available for recall from sale or supply of any batch or product should that become necessary.

GMP is often confused with *quality assurance* and *quality control*. Quality assurance encompasses all the organized arrangements made with the object of ensuring that the products will be fit for their intended use. It includes GMP and other factors such as original product design, product development, storage and distribution of the final product. Quality control is that part of GMP which is concerned with sampling and testing to ensure that the product meets specification.

Fig. 6.14 Protein inclusions in *E. coli* overproducing a recombinant protein. The inset shows normal *E. coli* cells. Photograph reproduced from *Science* courtesy of Dr D.C. Williams (Eli Lilly & Co.) and the American Association for the Advancement of Science.

Processes dependent on general microbial metabolism

The one microbiological process that most affects the quality of modern life, at least in the developed countries, is the disposal of industrial and domestic effluent. Without sewage treatment plants we would be awash with animal and human excreta and other noxious substances, disease would be more prevalent and we would have difficulty in obtaining water of suitable quality for human or industrial use. The disposal of effluent is a simple process and is shown diagrammatically in Fig. 6.15. Successful operation relies on the fact that sewage contains many different organic chemicals which can serve as substrates for microbial growth. The basic principle is the conversion of organic wastes to microbial cells and for maximum efficiency has to be an aerobic process. In the activated sludge process the sewage is vigorously agitated in large, open tanks. In the trickling filter process sewage is allowed to trickle down columns of coarse gravel through which air can permeate. Both processes are operated continuously. A number of newer designs are being introduced which are more efficient in terms of land use and energy consumption.

The digestion process used to treat the sludge resulting from the primary and secondary treatments is an anaerobic process. Its purpose is threefold: to reduce the volume of solids requiring disposal; to reduce the odour; and to reduce the number of pathogenic organisms. In some instances the methane produced during anaerobic digestion can be recovered and used to generate electricity. Ultimately the digested sludge is used for landfill or as a general-purpose fertilizer.

The objective of waste disposal systems is to degrade all the organic molecules in the effluent. However, during the past quarter of a century a number of man-made chemicals resistant to biodegradation have been released into the environment. Plastics and nylon are good examples and their persistence is due to the fact that they do not exist naturally and, having unusual chemical linkages, microorganisms do not exist which can degrade them. Of particular significance are com-

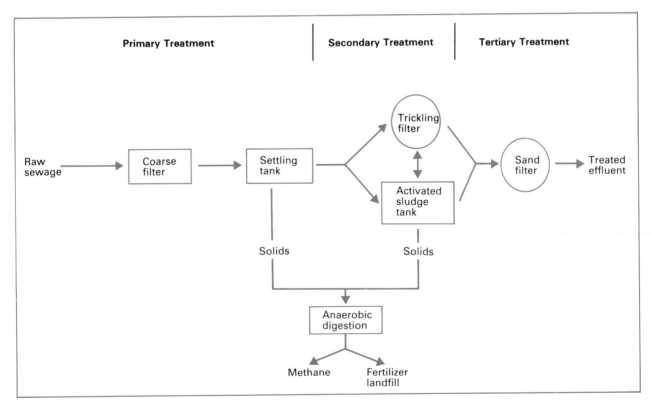

Fig. 6.15 Schematic representation of a sewage treatment works. In some plants effluent passes through both an activated sludge tank and a trickling filter.

pounds such as chlorinated hydrocarbons because, unlike plastics, they are not biologically inert but are cytotoxic, mutagenic and carcinogenic. Examples of such toxic chemicals are DDT and the herbicide 2,4,5-trichlorophenoxyacetic acid (2,4,5-T, Agent Orange).

Bacterial plasmids have been identified which carry genes specifying the degradation of chlorinated aromatic compounds such as 3-chlorobenzoate. Cells carrying these plasmids do not metabolize 4-chlorobenzoate unless a second plasmid encoding toluene and xylene decomposition is also present. By combining plasmids in this way it is possible greatly to expand the range of substrates which microbes can metabolize. After a range of microorganisms carrying different plasmids encoding degradation of various aromatic compounds were incubated with 2,4,5-T for 8−10 months an organism capable of growing on 2,4,5-T as sole carbon source was isolated. This organism contains a plasmid which has recruited genes from a variety of different plasmids. Although this genetic rearrangement occurred spontaneously, the process could be speeded up by the use of recombinant DNA technology. Based on a knowledge of the biochemistry of catabolism of aromatic compounds it should be possible to identify those genes which need to be united.

MICROBIOLOGICAL MINING [32, 33]

Microorganisms have been used in mineral leaching and metal concentration for many years. Indeed, more than 10% of the copper produced by the US is leached from ores by microorganisms. The microorganisms used are found naturally associated with ores; the ores are not deliberately inoculated. To date, the use of microorganisms in the mining industry has received relatively little attention because of the ease of mining high-grade ores. As high-grade ores are depleted and energy costs soar microbial leaching becomes more attractive. Currently leaching is restricted to copper and uranium ores but the commonest organism involved, *Thiobacillus ferrooxidans*, can also effect the solubilization of cobalt, nickel, zinc and lead.

The leaching of metals from ores can occur directly or indirectly. In the direct attack on metal sulphides the reaction which occurs is:

$$MS + 2O_2 \rightarrow MSO_4$$

where M can be Zn, Pb, Cu, Co, Ni, etc. In the indirect

Table 6.11 Desirable changes in *Thiobacillus ferrooxidans* strains used in microbiological mining.

1 An enhancement of the rate of regeneration of the ferric ion
2 Greater tolerance to acidic conditions
3 Greater tolerance to saline conditions
4 Increased resistance to metal ions such as Ag, Hg, Cd
5 Increased ability to withstand high temperatures

attack, pyrites (FeS_2) is oxidized to ferric sulphate and sulphuric acid and the ferric sulphate oxidizes the metal sulphide, e.g.:

$$4\,FeS_2 + 15O_2 + 2H_2O \rightarrow 2\,Fe_2(SO_4)_3 + 2H_2SO_4$$
$$CuS + Fe_2(SO_4)_3 \rightarrow CuSO_4 + 2FeSO_4 + S.$$

Whether direct or indirect attack occurs, the result is the same. The metal sulphate-containing solutions are collected and the metals removed by chemical precipitation.

Ways in which biotechnology could enhance the efficiency of microbial leaching are shown in Table 6.11. Of those listed, the easiest to achieve would be increasing the resistance to heavy metals. Unfortunately, *T. ferrooxidans* is a difficult organism to cultivate and thus not particularly amenable to genetic manipulation. But, where there is a will, there is a way!

Further reading

GENERAL REFERENCES

Bull A.T., Ellwood D.C. & Ratledge C. (1979) *Microbial Technology: Current State, Future Prospects. Symposium 29 of the Society for General Microbiology.* Cambridge University Press, Cambridge.
Peppler H.J. & Perlman D. (1979) *Microbial Technology,* vols I and II, 2nd edn. Academic Press, London.
Rehm H.−J. & Reed G. (1983 onwards) *Biotechnology, a Comprehensive Treatise,* vols II−VII. Springer-Verlag, Basle.
Each of these treatises contains one or more specific articles dealing at length with topics covered in this chapter.

CITED REFERENCES

1 Wolber P. & Warren G. (1989) Bacterial ice-nucleation proteins. *Trends in Biochemical Sciences* **14**, 179−82.
2 Aronson A.I., Beckmann W. & Dunn P. (1986) *Bacillus thuringiensis* and related insect pathogens. *Microbiological Reviews* **50**, 1−24.
3 Prince R.C. (1990) At least one *Bacillus thuringiensis* toxin forms ion-selective pores in membranes. *Trends in Biochemical Sciences* **15**, 2−3.
4 Honée G., Vriezen W. & Visser B. (1990) A translation fusion

product of two different insecticidal crystal protein genes of *Bacillus thuringiensis* exhibits an enlarged insecticidal spectrum. *Applied and Environmental Microbiology* **56**, 823–5.

5 Obukowitcz M.G., Perlak F.J., Kusano-Kretzmer F., Mayer E.J. & Wartrud L.S. (1986) Integration of the delta-endotoxin gene of *Bacillus thuringiensis* into the chromosome of root-colonizing strains of pseudomonads using Tn5. *Gene* **45**, 327–31.

6 Vedamuthu E.R. & Washam C. (1983) Cheese. In *Biotechnology*, vol. V, *Food and Food Production with Microorganisms*, ed. H.−J. Rehm & G. Reed, p. 231. Verlag-Chemie, Basle.

7 Primrose S.B. (1990) Controlling bacteriophage infections in industrial bioprocesses. *Advances in Biochemical Engineering/ Biotechnology* **43**, 1–10.

8 Daly T. & Fitzgerald G. (1987) Mechanisms of bacteriophage insensitivity in the lactic streptococci. In *Streptococcal Genetics* ed. J.J. Ferretti & R. Curtiss, p. 259. American Society for Microbiology, Washington.

9 Vasey R.B. & Powell K.A. (1984) Single-cell protein. *Biotechnology and Genetic Engineering Reviews* **2**, 285–311.

10 Kaper J.B., Mobley H.L.T., Michalski J.M., Herrington D.A. & Levine M.M. (1988) In *Advances in Research on Cholera and Related Diarrhoeas*, vol. VI, ed. N. Ohtomo & R.B. Sack, p. 161. KTK Scientific Publishers, Tokyo.

11 Makoff A.J., Ballantine S.P., Smallwood A.E. & Fairweather N.F. (1989) Expression of tetanus toxin fragment C in *E. coli*: its purification and potential use as a vaccine. *Bio/technology* **7**, 1043–6.

12 Charles I. & Dougan G. (1990) Gene expression and the development of live enteric vaccines. *Trends in Biotechnology* **8**, 117–21.

13 Curtiss R. (1990) Antigen delivery systems for analysing host immune responses and for vaccine development. *Trends in Biotechnology* **8**. 237–40.

14 Hodgson J. (1990) UK sweet on engineered yeast. *Bio/technology* **8**, 281.

15 Chater K.F. (1990) The improving prospects for yield increase by genetic engineering in antibiotic-producing streptomycetes. *Bio/technology* **8**, 115–21.

16 Smith D.J., Burnham M.K.R., Edwards J., Earl A.J. & Turner G. (1990) Cloning and heterologous expression of the penicillin biosynthetic gene cluster from *Penicillium chrysogenum*. *Bio/technology* **8**, 39–41.

17 Hopwood D.A., Malpartida F., Kieser H.M., Iheda H., Duncan J., Fujii I., Rudd B.A.M., Floss H.G. & Omura S. (1985) Production of 'hybrid' antibiotics by genetic engineering. *Nature* **314**, 642–4.

18 Baldwin J.E. & Abraham E. (1988) The biosynthesis of peni-

cillins and cephalosporins. *Natural Products Reviews* **5**, 129–145.

19 Anderson S., Marks C.B., Lazarus R., Miller J., Stafford K., Seymow J., Light D., Rastetter W. & Estell D. (1985) Production of 2-*keto*-L-gulonate, an intermediate in L-ascorbate synthesis, by a genetically modified *Erwinia herbicola*. *Science* **230**, 144–9.

20 Ensley B.D., Ratzkin B.J., Osslund T.D., Simon M.J., Wackett L.P. & Gibson D.T. (1983) Expression of napthalene oxidation genes in *Escherichia coli* results in biosynthesis of indigo. *Science* **222**, 167–9.

21 Green M.L., Angal S., Lowe P.A. & Marston F.A.O. (1985) Cheddar cheesemaking with recombinant calf chymosin synthesized in *Escherichia coli*. *Journal of Dairy Research* **52**, 281–6.

22 Arbige M.V. & Pitcher W.H. (1989) Industrial enzymology: a look towards the future. *Trends in Biotechnology* **7**, 330–5.

23 Van Brunt J. & Klawner A. (1988) Growth factors speed wound healing. *Bio/technology* **6**, 25–30.

24 Horwich A.L., Newpert W. & Hartl F-U. (1990) Protein-catalyzed protein folding. *Trends in Biotechnology* **8**, 126–31.

25 Schein C.H. (1990) Solubility as a function of protein structure and solvent components. *Bio/technology* **8**, 308–17.

26 Marullo S., Delavier-Klutchko C., Guillet J-G., Charbit A., Strosberg A.D. & Emorine L.J. (1989) Expression of human β1 and β2 adrenergic receptors in *E. coli* as a new tool for ligand screening. *Bio/technology* **7**, 923–7.

27 Goodson R.J. & Katre N.V. (1990) Site directed pegylation of recombinant interleukin-2 at its glycosylation site. *Bio/technology* **8**, 343–6.

28 Parekh R.B., Dwek R.A., Edge C.J. & Rademacher T.W. (1989) *N*-glycosylation and the production of recombinant glycoproteins. *Trends in Biotechnology* **7**, 117–22.

29 Gatenby A.A., Viitanen P.V. & Lorimer G.H. (1990) Chaperonin assisted polypeptide folding and assembly: implications for the production of functional proteins in bacteria. *Trends in Biotechnology* **8**, 354–8.

30 Ohlsson K., Bjork P., Bergenfeldt M., Hageman R., & Thompson R.C. (1990) Interleukin-1 receptor antagonist reduces mortality from endotoxin shock. *Nature* **348**, 550–2.

31 Anderson A.J. & Dawes E.A. (1990) Occurrence, metabolism, metabolic role and industrial uses of bacterial polyhydroxyalkanoates. *Microbiological Reviews* **54**, 450–72.

32 Mergeay M. (1991) Towards an understanding of the genetics of bacterial metal resistance. *Trends in Biotechnology* **9**, 17–24.

33 Lundgren D.G. & Malouf E.E. (1983) Microbial extraction and concentration of metals. *Advances in Biotechnological Processes* **1**, 223–49.

7 Cell and Enzyme Immobilization

Microorganisms act as a catalyst in any fermentation where a product other than biomass is obtained. When the culture medium is removed from the fermenter the catalyst is either destroyed by subsequent downstream processing or discarded. It is possible to harvest cells and reuse them but there are a number of practical problems. Thus enzyme activity is lost unless there is a further period of cell growth and contamination is a major problem. One solution is to immobilize the microorganisms or even the appropriate enzyme.

The use of immobilized cells is not a new idea. Since the early eighteenth century they have been used in the production of vinegar and since the early part of this century for disposal of sewage effluent. In the vinegar process alcohol is trickled down a column of beechwood shavings which are coated with a film of acetic acid bacteria that carry out the reaction:

$$CH_3CH_2OH \longrightarrow CH_3CHO \longrightarrow CH_3COOH.$$

$$NAD^+ \ NADH \quad NAD^+ \ NADH$$
$$+ \ H^+ \qquad\qquad + \ H^+$$

A similar technique is used for sewage effluent disposal except that sand and gravel act as the microbial support and the desired end-result is conversion of all the organic materials in the effluent to cells, water, carbon dioxide and other gases. In both instances the microbial film forms naturally and is held in place by a combination of microbially produced exopolysaccharides ('slime') and physical forces. Since growth of the organisms occurs, enzyme decay is not a problem. There is also a continual release of aggregates of microorganisms ('flocs').

Although these processes for vinegar manufacture and sewage disposal are perfectly satisfactory, it would be preferable if in other processes using immobilized cells the effluent stream were free of microorganisms. In order to achieve this the cells must be firmly bonded to the support matrix or trapped in some other way and cell growth must not occur. The latter demands that the requisite enzymes have long half-lives. As it turns out, these criteria can be met.

IMMOBILIZATION METHODS

The methods of immobilization available are equally applicable to cells and enzymes. Three categories of method can be discerned: physical binding, cross-linking and entrapment (Fig. 7.1). In practice, combinations of methods may be used. Physical binding is the oldest immobilization technique but the least satisfactory. Cells or enzyme are mixed with adsorbent and then packed in a column. However, ease of adsorption also means ease of desorption and this often occurs following substrate addition. Cells immobilized in this way have a tendency to autolyse and if enzymes are immobilized activity is often partially or totally lost.

Enzymes and microbial cells can be immobilized by cross-linking them with bi- or multifunctional reagents such as glutaraldehyde or toluene diisocyanate. They can also be bonded to insoluble matrices using the same reagents. Although good retention of cells and enzymes is obtained with this method, it can be accompanied by extensive loss of enzyme activity.

The most extensively used method of cell immobilization involves entrapment in a polymer matrix. Matrices which have been employed include collagen, gelatin, agar, alginate, carrageenan, polyacrylamide, cellulose triacetate and polystyrene. For immobilization with acrylamide, cells are mixed with acrylamide monomer, a polymerizing agent such as N,N'-methylene bisacrylamide, potassium persulphate to initiate polymerization and β-dimethylaminopropionitrile, which is an accelerator of polymerization. After 30–60 min at ambient temperatures a stiff gel forms and this can be granulated to a suitable size, packed in a column and washed with saline to remove residual chemicals. Carrageenan is in some ways an even more advantageous matrix for immobilization: the procedure is much simpler and uses no noxious chemicals so that there is better retention of enzyme activity. Cells are suspended in saline at 45–50 °C and carrageenan is dissolved in saline at a similar temperature. The two solutions are mixed and cooled to 10 °C. To increase

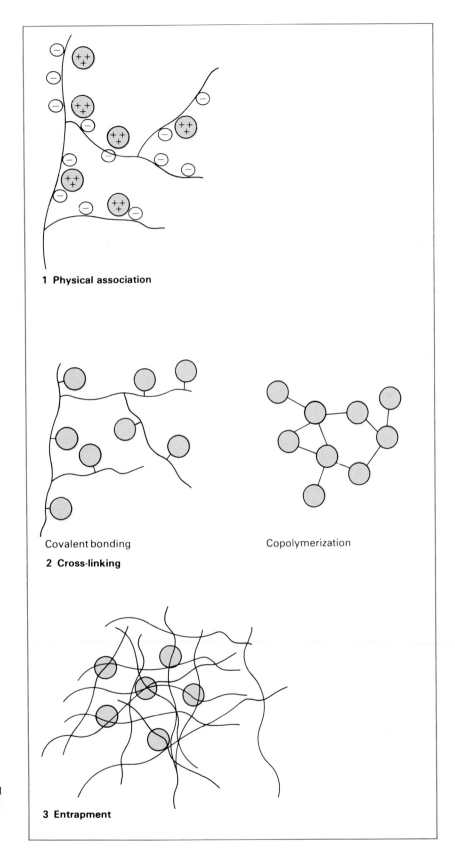

1 Physical association

Covalent bonding   Copolymerization

2 Cross-linking

3 Entrapment

Fig. 7.1 The three basic methods of cell and enzyme immobilization. The tinted circles represent enzyme molecules or microbial cells.

the gel strength it is soaked in cold potassium chloride. After this treatment the gel is granulated to a suitable size. If the operational stability of the support matrix is not satisfactory, then a hardening agent such as glutaraldehyde or hexamethylenediamine can be added.

CHOICE OF IMMOBILIZATION METHOD [1]

The art of immobilization lies in knowing which method to choose. This, in turn, is governed by a number of factors, some of which will not be apparent until the procedure is tried. At least five factors are likely to be of paramount importance:

1 the operational stability of the immobilized catalyst;
2 the cost of the immobilized catalyst;
3 the activity and yield of the immobilized catalyst;
4 the regenerability of the catalyst;
5 the cost of the appropriate reactor configuration.

Initially the choice of method is empirical but subsequently the process must be thoroughly analysed and tested and considered on its merits.

EXAMPLE: IMMOBILIZATION OF L-AMINO ACID ACYLASE

There is a large market for L-amino acids as food and feed supplements. Some L-amino acids can be made efficiently by fermentation; others cannot and have to be prepared chemically. Although chemical synthesis is often easier and cheaper, the immediate products are usually a mixture of the D- and L-isomers and resolution is necessary. One method of resolution is to acylate the amino acid (Fig. 7.2) and selectively hydrolyse the L-acyl derivative with the enzyme aminoacylase. The L-amino acid and the D-acyl amino acid can be separated easily by virtue of their differing solubilities. The D-acyl amino acid can be discarded

or, better, racemized to the D,L-form and reused.

Industrial processes for L-amino acid production based on the use of soluble aminoacylase were introduced over 30 years ago. These processes suffered from two basic drawbacks: complicated separation of product and enzyme with concomitant low yields and high costs, and non-reusability of the enzyme. To overcome these problems immobilization techniques were developed in Japan by Chibata and Tosa and in 1969 aminoacylase became the first immobilized enzyme to be used commercially on a large scale (Fig. 7.3).

Initially over 40 different immobilization procedures were tried but most were discarded for one or more reasons, e.g. instability of the support matrix, low yield and/or low activity of the preparation. The three most promising methods which emerged were ionic binding to DEAE-Sephadex, covalent bonding to iodoacetyl cellulose and entrapment within polyacrylamide. A detailed comparison of the operational properties of the enzyme immobilized by those three methods is shown in Table 7.1. It is worth noting that some of the kinetic properties of the immobilized enzyme preparations are quite different from those of the free enzyme; this is a common observation and frequently there is no obvious explanation. From the data in Table 7.1 it can be seen that there is little to choose between the performances of the various preparations, with the possible exception of the long half-life of the aminoacylase immobilized on DEAE-Sephadex. However, once cost factors and ease of preparation are included in the assessment (Table 7.2) selection becomes easier. A key feature of the use of DEAE-Sephadex for immobilization is that it can be recycled when the activity of the preparation falls below the desired level, and this led to the choice of the DEAE-Sephadex-immobilized aminoacylase for commercial development. In practice, the activity of

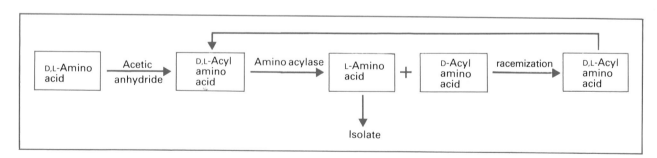

Fig. 7.2 Resolution of amino acid racemates by acylation and selective (enzymic) deacylation.

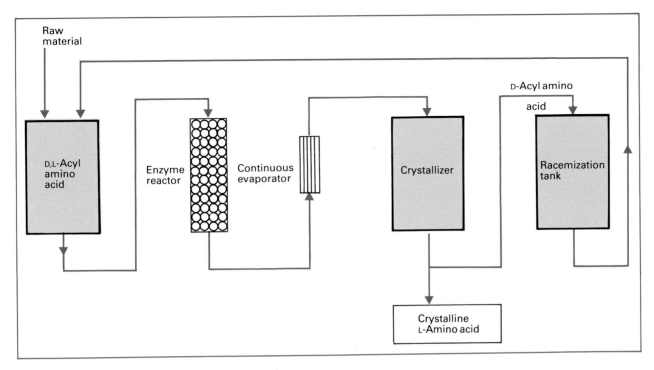

Fig. 7.3 Schematic representation of the commercial process for amino acid resolution.

Table 7.1 Comparison of properties of free and immobilized aminoacylase.

Physicochemical property	Free aminoacylase	Method of immobilizing aminoacylase		
		DEAE-Sephadex	Iodoacetyl cellulose	Polyacrylamide entrapment
pH optimum	7.5−8.0	7.0	7.5−8.0	7.0
Temperature optimum	60°C	72°C	55°C	65°C
Activation energy at 37°C and pH 7.0	6.9 kcal/mol	7.0 kcal/mol	3.9 kcal/mol	5.3 kcal/mol
K_m at 37°C and pH 7.0	5.7 mmol/l	8.7 mmol/l	6.7 mmol/l	5.0 mmol/l
V_{max} at 37°C and pH 7.0	1.52 mol/h	3.33 mol/h	4.65 mol/h	2.33 mol/h
% Retention of activity after 10 min at 70°C	12.5	87.5	62.5	34.5
Half-life	−	65 days at 50 °C	−	48 days at 37°C

an enzyme column drops by 40% during 30 days' operation at 50 °C but is regenerated by the addition of fresh enzyme.

A detailed analysis of the recurrent costs associated with the immobilized aminoacylase system has shown them to be more than 40% less than those incurred with the original soluble enzyme process (Fig. 7.4). As would be expected, considerable savings can be made because the enzyme is used for long periods and is not discarded and, because the immobilized enzyme process is continuous, it can be automated and this, in turn, reduces labour costs. The savings in substrate cost are due to the ease of recovery of product which results in higher yields.

OTHER APPLICATIONS OF IMMOBILIZED CATALYSTS [2, 3, 4, 5]

Apart from the resolution of D,L-amino acids, immobilized catalysts are used for the commercial production

Table 7.2 Factors to be taken into account in determining method of choice for immobilization of aminoacylase.

Factor	Method of immobilization of aminoacylase		
	DEAE-Sephadex	Iodoacetyl cellulose	Polyacrylamide entrapment
Ease of preparation	Easy	Difficult	Intermediate
Retention of activity on immobilization	High	High	High
Cost of immobilization	Low	High	Moderate
Strength of binding	Medium	Strong	Strong
Operational stability (half-life)	Long	—	Moderate
Ability to be regenerated	Possible	Not possible	Not possible

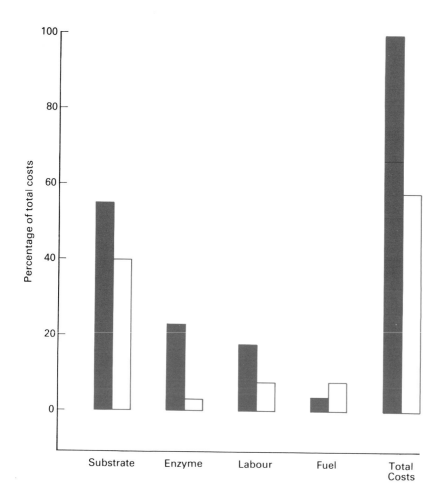

Fig. 7.4 Comparison of costs associated with soluble (solid red bars) and immobilized (open red bars) enzyme processes for aminoacylase resolution of D,L-amino acids.

of a number of other chemicals (Table 7.3). Among these is the best example of the commercial realization of immobilized biocatalyst technology, the production of high-fructose corn syrup (HFCS). The production of HFCS involves the use of two amylases and glucose isomerase. The amylases effect the liquefaction and subsequent degradation (*saccharification*, meaning sugar formation) of the cornstarch to glucose. The glucose isomerase then converts the glucose to fructose until equilibrium is reached when the two are approxi-

Table 7.3 Commercial application of immobilized cells and enzymes.

Product	Use	Enzyme and reaction	Catalyst
L-Amino acids	Various	aminoacylase RCHCOOH $\longrightarrow$ R-CHCOOH + RCHCOOH |　　　　　　　|　　　　| NH-COR′　　　　NH$_2$　　NH-COR′ D,L-acyl amino acid　　L-amino acid　D-acyl amino acid	Enzyme immobilized on DEAE-Sephadex
High-fructose syrup	Fructose approx. 1.5 times sweeter than sucrose	glucose 100% glucose syrup $\longrightarrow$ 42% fructose (from starch)　isomerase　50% glucose 8% other sugars	Enzyme immobilized on cellulose ion-exchanger
6-Amino-penicillanic acid	Chemical modification to produce semi-synthetic penicillins	penicillin R-CONH— ... $\longrightarrow$ H$_2$N ... amidase naturally produced penicillins → 6-amino penicillanic acid	Enzyme immobilized on sepharose with cyanogen bromide or *E. coli* cells entrapped in polyacrylamide gels
Urocanic acid	Sunscreen for use in suntan lotion	histidine CH$_2$CHN H$_2$COOH $\longrightarrow$ CH·CH COOH ammonia lyase histidine → urocanic acid	*Achromobacter liquidum* immobilized in polyacrylamide
Malic acid	Acidulant in food industry	COOH　　　　　COOH |　　　　　　　| CH　fumarase　CHOH || $\longrightarrow$ | CH　　　　　　CH$_2$ |　　　　　　　| COOH　　　　　COOH fumaric acid　　malic acid	*Brevibacterium flavium* immobilized in carrageenan
Aspartic acid	Production of the artificial sweetener aspartame	COOH　　　　　　COOH |　　　　　　　　| CH　aspartase　CHNH$_2$ || $\longrightarrow$ | CH　+ NH$_3$　CH$_2$ |　　　　　　　　| COOH　　　　　　COOH fumaric acid　　aspartic acid	*E. coli* cells immobilized in carrageenan

mately equimolar. The interest in HFCS stems from the fact that fructose is sweeter than glucose yet has the same calorific value; the inclusion of HFCS instead of glucose syrup in processed foods permits the same level of sweetness but provides fewer calories. In the period 1970–1980 HFCS increased its share of US per capita consumption of nutritive (i.e. calorie-generating) sweeteners from almost nil to over 16%. In the same period sucrose usage decreased from 84% to 68%. By 1981 over 2.5 million metric tons (dry weight) of HFCS were produced.

The increasing demand for low-calorie sweeteners is responsible for the rapid development of another immobilized biocatalyst, that involved in the production of L-aspartic acid. The interest in aspartic acid stems from the fact that it is an ingredient of aspartame (aspartyl phenylalanine methyl ester). Since it is 160 times sweeter than sucrose only trace amounts of aspartame need to be added to food and thus it is considered to be non-nutritive, i.e. it contributes little to the calorific content of the food. In 1981 the US Food and Drug Administration approved the use of aspartame as a food additive. Since then sales of aspartame, and hence aspartic acid, have grown rapidly to thousands of tonnes per year. Indeed, the spectacular increase in the use of aspartame may eventually lead to a decline in the use of HFCS!

Another important chemical produced with the aid of an immobilized biocatalyst is 6-aminopenicillanic acid. This is a key intermediate in the production of the semi-synthetic penicillins (Fig. 7.5), which are the mainstay of current antimicrobial chemotherapy. The original fermentations of *Penicillium chrysogenum* (see p. 6) produced a mixture of penicillins but by appropriate medium supplementation could be induced to synthesize benzylpenicillin almost exclusively. Benzylpenicillin revolutionized the treatment of many potentially lethal bacterial infections but in spite of this was a less than perfect antimicrobial agent. In particular, it exhibited a restricted antibacterial spectrum, it had to be administered by injection since it was broken down by gastric acidity when given orally, and it was hydrolysed by enzymes known as β-lactamases which are produced by many bacteria.

A quite different application of immobilization, which is still in its infancy, is the manufacture of enzyme electrodes. These are probes which are capable of generating an electrical potential as a result of a reaction catalysed by an immobilized enzyme that is fixed on to or around the probe. The general principle of enzyme electrodes is embodied in the glucose electrode. In it, glucose oxidase is immobilized in a polyacrylamide gel which surrounds an oxygen electrode. The operating principle is that the enzyme catalyses the removal of oxygen from solution at a rate proportional to the concentration of glucose present.

Fig. 7.5 The synthesis of improved penicillins by chemical modification of 6-aminopenicillanic acid produced from benzylpenicillin by immobilized *E. coli* cells containing penicillin amidase.

A whole series of enzyme electrodes have been constructed but to be of practical value they must be easy to make, cheap, stable, sensitive and have a low response time. Electrodes which fulfil these criteria are available for many of the commercially important microbial metabolites, e.g. penicillin and various amino acids, and, in theory, could be invaluable for on-line analysis of fermentation broths (see p. 67). In practice, they are not robust enough, nor can they be sterilized.

THE CHOICE BETWEEN IMMOBILIZED CELLS AND IMMOBILIZED ENZYMES

In some instances there is no difficulty in choosing between cells or enzymes for immobilization. For example, if the enzyme with the desired activity is secreted from the cell, then clearly it is the enzyme which should be immobilized. With protein engineering it is even possible to combine several enzymic activities in a single protein molecule. If activity of the enzyme is dependent on its spatial arrangement in the cell membrane, then clearly it is the cell which should be immobilized. More often there is little to choose between the two systems and then the overriding factor is cost. In this context it is not the cost of immobilization but the sum of the capital and recurrent costs associated with initiating and running the process. Of particular importance are the retention of activity on immobilization and the operational half-life.

To date the commercially developed immobilized systems (Table 7.3) are ones in which the catalyst has no requirement for a cofactor. The commonest cofactors are $NAD^+/NADP^+$ and ATP and their cost is such that it is not economically feasible to add them continuously with the substrate. The only practical proposition is to recycle the cofactors. This can be done in enzyme reactors (Fig. 7.6) but usually two substrates must be added and two products are obtained, only one of which is required. In addition, the cofactor also must be immobilized or it will be lost from the system and such immobilized cofactors tend to have a short half-life. Immobilized cells may be more useful in this respect for, if the cell membrane remains intact during the immobilization step, the coenzymes are retained as are all the enzymes necessary for recycling them. However, if the cell membrane does remain intact, then diffusion of substrate and product across it may be rate-limiting.

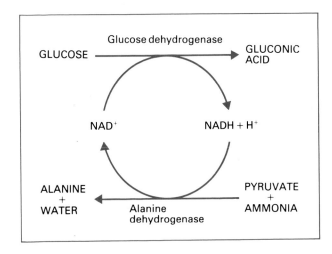

Fig. 7.6 A coupled enzyme reaction for the recyling of a pyridine nucleotide cofactor.

SOME OPERATIONAL CONSIDERATIONS

A number of factors can affect the operational stability of immobilized cell and enzyme reactors. One of these is the purity of the substrate used. Impurities are important for two reasons. First, they can accumulate during prolonged reactor operation and either foul the reactor bed or, worse still, irreversibly inactivate the catalyst. Second, these impurities can act as substrates for growth of microbial contaminants. Growth of contaminants can be minimized by the use, wherever possible, of high temperatures and pH values and high substrate concentrations.

A different problem is gas exchange. In some immobilized cell systems the viability of the cells has to be maintained and this can mean providing a supply of oxygen. This can best be achieved by using a fluidized bed. In such a system the beads of immobilized cells are kept in suspension by the introduction of air into the bottom of the reactor. Clearly a design has to be selected in which the immobilized cells can be prevented from leaving the reactor in the effluent stream. An alternative situation is one in which a gas such as carbon dioxide is a product of the reaction. Here the gas can be stripped out of the reactor by using high substrate feed rates (*medium velocity*).

A factor which can preclude the use of immobilized biocatalysts is the generation of heat. It is not easy to design reactors for immobilized cells and enzymes which efficiently remove the heat produced by an exothermic reaction. Thus the heat generated in the middle of an aggregate of immobilized catalyst may be sufficient to inactivate the enzyme. Since the heat produced will be proportional to the substrate concentration, one solution is to reduce the concentration of substrate in the feed solution. However, this will reduce the concentration of product in the reactor effluent making product recovery more difficult and expensive and the process may not be economic. It should be noted that the generation of heat might not be observed on the laboratory scale because the surface area to volume ratio of the reactor is such that the heat produced is dissipated by conduction and convection. Thus before embarking on construction of a large-scale reactor it is advisable to check the thermodynamics of the reaction in a calorimeter.

ROLE OF RECOMBINANT DNA TECHNOLOGY

In many instances biological processes have not been developed because of the high cost of the catalyst. Thus, enzymic hydrolysis of lactose has not been economically viable because of the costs associated with production of the enzyme β-galactosidase. In this respect recombinant DNA technology is of particular value since it is an easy matter to increase the level of an enzyme in the cell from less than 0.1% to over 20% of total cell protein. Even where a non-recombinant process is viable, increasing the enzyme level in this way should cut catalyst costs and permit greater catalyst loadings per unit volume of reactor. As indicated above, an important feature of an immobilized catalyst is its half-life. Gene manipulation can be of value in this context for, as discussed in Chapter 4, it is possible

to engineer proteins such that their *in vitro* stability is increased, e.g. by introducing additional disulphide bridges.

Further reading

GENERAL REFERENCES

Chibata I., Tosa T. & Sato T. (1979) Use of immobilized cell systems to prepare fine chemicals. In *Microbial Technology*, 2nd edn, ed. H.J. Peppler & D. Perlman, Vol. II, pp. 433–61. Academic Press, London.

Klibanov A.M. (1983) Immobilized enzymes and cells as practical catalysts. *Science* **219**, 722–7.

Laskin A.I. (ed.) (1985) *Enzymes and Immobilized Cells in Biotechnology*. Benjamin Cummings, Menlo Park.

Trevan M.D. (1980) *Immobilized Enzymes*. J. Wiley & Sons, Chichester.

CITED REFERENCES

1 Wood L.L. & Calton G.J. (1984) A novel method of immobilization and its use in aspartic acid production. *Bio/technology* **2**, 1081–4.

2 Jensen S.E., Westlake D.W.S. & Wolfe S. (1984) Production of penicillins and cephalosporins in an immobilized enzyme reactor. *Applied Microbiology and Biotechnology* **20**, 155–60.

3 Klein M.D. & Langer R. (1986) Immobilized enzymes in clinical medicine: an emerging approach to new drug therapies. *Trends in Biotechnology* **4**, 179–86.

4 Kricka L.J. & Thorpe G.H.G. (1986) Immobilized enzymes in analysis. *Trends in Biotechnology* **4**, 253–8.

5 Vandamme E.J. (1983) Peptide antibiotic production through immobilized biocatalyst technology. *Enzyme and Microbial Technology* **5**, 403–15.

Part IV
Animal Biotechnology

8 Principles of Mammalian Cell Culture

The cultivation of animal tissue *in vitro* was first shown to be possible in 1907 but the reproducible and reliable large-scale culture of mammalian cells has been achieved only in the last 10–15 years. Originally large-scale culture was developed for the production of animal viruses to be used as vaccines. More recently the goal has been the production of human proteins of potential therapeutic value and monoclonal antibodies; these topics are discussed in the next two chapters.

ESTABLISHING CELLS IN CULTURE

The cultivation of mammalian cells from a particular tissue begins with the dissociation of the tissue fragment into its component cells by treatment with a proteolytic enzyme, usually trypsin. After removal of the trypsin the cell suspension is placed in a flat-bottomed glass or plastic container together with a suitable liquid medium. After a lag period the cells attach themselves to the bottom of the container and start dividing mitotically. A culture of this type, arising directly from differentiated tissue, is referred to as a *primary culture*. Eventually the bottom of the culture vessel will be covered with a continuous layer of cells, often one cell thick and hence referred to as a *monolayer*.

Primary cultures are usually prepared from relatively large tissue masses and so will comprise a variety of differentiated cells, e.g. fibroblasts, macrophages, epithelial cells, lymphocytes, etc. However, the cells that multiply best in culture have the spindle shape (Fig. 8.1(a)) and growth rate of connective tissue cells and are called *fibroblasts*. Since epithelial cells possess the enzyme D-amino acid oxidase whereas fibroblasts do not, it is possible selectively to promote the growth of epithelial cells in culture by using D-valine as the sole source of valine. Such epithelial cells are polygonal in shape (Fig. 8.1(b)). However, the shape of cells in culture is not invariable and can be affected by the composition of the medium and the presence of infectious agents.

The cells of primary cultures can be detached from the culture vessel by trypsin treatment or the addition of the chelating agent EDTA. These cells can be used to initiate *secondary cultures* by reseeding them in fresh media at high density. Cells from primary cultures can often be transferred serially a number of times. The cells multiply at a constant rate over successive transfers and such cells comprise a *cell strain*. Cell strains do not have infinite life and divide only a finite number of times before their growth rate declines and they die; for example, human cells generally divide only 50–100 times before dying.

CONTINUOUS CELL LINES

Not all animals yield primary cultures all of whose secondary progeny die. Most murine cells die after dividing 30–50 times but a few cells become altered such that they acquire a different morphology, grow faster and are able to start a culture from a smaller number of cells. Progeny derived from such exceptional cells have unlimited life, unlike the cell strains from which they originated, and are designated *cell lines*.

During repeated serial transfer cell lines can undergo extensive changes in their cultural properties; for example, the density at which division ceases may increase such that the cells grow in clumps rather than in monolayers and the cells may be irregularly oriented in respect to each other. Such cell lines are said to be *transformed* and are generally *neoplastic*, that is they produce cancer if transplanted into related animals. Lines of transformed cells can also be obtained from normal primary cell cultures or cell strains by infecting them with oncogenic viruses or treating them with carcinogenic chemicals. Whereas no human cell line has been shown unambiguously to have originated from normal cells a number have been derived from human tumours. Apparently the possession of the cancerous phenotype allows easier adaptation to growth in cell culture. This may be related to the fact that cancer cells, unlike those of cell lines derived from primary cultures, generally possess extra chromosomes, that is, they are *aneuploid*. The properties of a

(a)

(b)

Fig. 8.1 Morphology of human cells in culture. (a) Fibroblasts; (b) epithelial cells.

number of the common cell lines are shown in Table 8.1.

The tendency of cell lines to change continuously on repeated cultivation necessitates that stocks of cells generally be maintained in the frozen state. The cells are mixed with additives such as glycerol or dimethyl sulphoxide to minimize cellular damage by ice crystals and are dispensed in ampoules and stored in liquid nitrogen. Cells maintained in this way remain viable for years and on thawing readily initiate new cultures.

Just as the cultural properties of cell lines may be different from those of normal cells, so too is their metabolism; for example, cells from liver-derived hepatomas fail to synthesize most of the normal liver enzymes but may continue to secrete the major liver product albumin. Similarly, cell lines from some tumours secrete hormones in an uncontrolled fashion, e.g. cells from tumours of the testes secrete androgen while pituitary tumour cell lines secrete adrenocorticotrophic hormone and growth hormones.

Table 8.1 Properties of some commonly used mammalian cell lines.

Cell line	Species of origin	Tissue of origin	Morphology	Ploidy	Growth in suspension
3T3	Mouse	Connective tissue	Fibroblast	Aneuploid	No
L	Mouse	Connective tissue	Fibroblast?	Aneuploid	Yes
CHO	Chinese hamster	Ovary	Epithelial	Quasidiploid	Yes
BHK21	Syrian hamster	Kidney	Fibroblast	Diploid	Yes
HeLa	Human	Cervical carcinoma	Epithelial	Aneuploid	Yes
WISH	Human	Amnion	Epithelial	Aneuploid	?
Hep-2	Human	Carcinoma of larynx	Epithelial	Aneuploid	?
KB	Human	Nasopharyngeal carcinoma	Epithelial	Aneuploid	Yes

MEDIA FOR MAMMALIAN CELL CULTURE

One of the most crucial factors in achieving the succssful cultivation of mammalian cells *in vitro* is the composition of the growth medium. To be satisfactory the medium must fulfil the criteria set out in Table 8.2.

The growth media in common use are complex and generally not completely defined because of the presence of serum, although this can sometimes be replaced simply with albumin. The basic component of these media is a balanced salt solution whose functions are to provide:

1 essential inorganic ions;
2 the correct osmolality;
3 the correct pH;
4 a source of energy in the form of glucose;
5 a pH indicator, phenol red.

The compositions of two widely used balanced salt solutions are shown in Table 8.3.

Although cells will remain alive for several hours in a balanced salt solution, many more ingredients must

Table 8.2 Criteria for a successful culture medium for animal cells.

1 The medium must provide all nutritional requirements of the cell
2 The medium must maintain a pH value of 7.0–7.3 despite the production of acid, i.e. it must be adequately buffered
3 The medium must be isotonic with the cell cytoplasm
4 The medium must be sterile

Table 8.3 Comparison of two balanced salt solutions.

	Amount (mg/l) in balanced salt solution of	
Ingredient	Earle	Hanks
NaCl	6800	8000
KCl	400	400
$CaCl_2 \cdot 2H_2O$	264	185
$MgSO_4 \cdot 7H_2O$	200	200
$NaH_2PO_4 \cdot H_2O$	140	—
Na_2HPO_4	—	47.5
KH_2PO_4	—	60
$NaHCO_3$	1680	350
Glucose	1000	1000
Phenol red	17	17

be included if proliferation of cells is to occur. These ingredients include most of the common amino acids, eight vitamins (Table 8.4) and 5–10% of serum. The precise contribution made by the serum is not fully understood. Some components of the serum, e.g. α-globulins and hormones, may exert a beneficial effect by promoting the attachment and spreading of the cells and by stimulating cell division. The most popular source of serum is from fetal calves since this maintains better growth of cultured cells than serum from adult animals. Not surprisingly, fetal calf serum is prohibitively expensive and generally in short supply. Serum must be obtained from a reputable supplier who has a rigorous quality control programme (Table 8.5). A few cell lines can be maintained in media

Table 8.4 Defined supplements added to balanced salt solutions for growth of HeLa cells. Figures in parentheses are for optimal growth of mouse fibroblasts.

Ingredient	Amount (mmol/l)
L-arginine HCl	0.1
L-cysteine 2HCl	0.05 (0.02)
L-glutamine	2.0 (1.0)
L-histidine HCl.H$_2$O	0.05 (0.02)
L-isoleucine	0.2
L-leucine	0.2 (0.1)
L-lysine HCl	0.2 (0.1)
L-methionine	0.05
L-phenylalanine	0.1 (0.05)
L-threonine	0.2 (0.1)
L-tryptophan	0.02 (0.01)
L-tyrosine	0.1
L-valine	0.2 (0.1)
D-calcium pantothenate	10^{-3}
Choline chloride	10^{-3}
Folic acid	10^{-3}
i-Inositol	10^{-3}
Nicotinamide	10^{-3}
Pyridoxal. HCl	10^{-3}
Riboflavin	10^{-4}
Thiamin. HCl	10^{-3}

Table 8.5 Some of the quality control tests carried out on high-quality serum for use in mammalian cell culture.

Test	Desired result
Sterility	No growth on a variety of media (e.g. blood agar, thioglycollate broth, Sabouraud's dextrose slant) at room temperature or 37°C
Mycoplasmas and viruses	Absent
Efficacy of cell growth	Compare cell growth in T-flask with control serum lot
Total protein content and immunoglobulin electrophoresis pattern	Same as content/pattern of control serum
pH value after filtration	7.2–7.8
Osmolality	290–310 mosm/l
Endotoxin content	Less than 2 ng/ml by the limulus amoebocyte lysate test
Haemoglobin content	Less than 20 mg%

sociates in solution, releasing carbon dioxide into the atmosphere and hydroxyl ions into the medium. If dissociation occurs faster than acid is produced by the cells, then the medium will turn alkaline. Thus the buffering capacity of the culture medium is better maintained if the proportion of the CO_2 in the gas phase inside the culture vessel is high. This is reflected in the differences in bicarbonate content of the two balanced salt solutions shown in Table 8.3. Cells which release a lot of metabolic CO_2 are best cultured in supplemented Hanks' saline in a sealed container where the CO_2 which they release helps slow down the dissociation of the sodium bicarbonate. Cells which do not produce much CO_2 do better in Earle's saline exposed to a gas mixture of 5% CO_2 in air. Because of the difficulties associated with the use of bicarbonate buffers a number of organic buffers have been introduced for animal cell culture. The most popular of these is HEPES (N-2-hydroxyethylpiperazine-N'-2-ethanesulphonic acid), which has a pK_a of 7.3 at 37 °C. Unfortunately this compound is too expensive for routine large-scale use.

The media used to culture animal cells are rich in nutrients so they are particularly susceptible to contamination with bacteria and fungi unless good aseptic technique is practised. For this reason antibiotics are usually added: penicillin to prevent growth of Gram-negative cells and nystatin to inhibit fungi. Another problem with cell cultures is contamination with mycoplasmas, a group of bacteria which lack cell walls. Most of the contaminating mycoplasmas are non-pathogenic and probably originate from the oropharynx of laboratory staff, but some are derived from trypsin or serum. Tetracyclines, kanamycin and gentamycin can be used to inhibit most mycoplasmas.

PREPARATION OF MEDIA AND EQUIPMENT FOR CELL CULTURE

Probably the most important aspect of animal cell culture is the quality of the water used. High-quality distilled water is required, and this is usually prepared from deionized water. The water used for washing and rinsing of equipment must be of the same quality, as should be the steam used for sterilization. Common laboratory or plant steam cannot be used since it contains additives toxic to mammalian cells and is contaminated with metal ions unless stainless steel is used throughout.

All the materials which come into contact directly or

devoid of any natural components but these must be considered the exception rather than the rule. More success has been achieved when the serum has been replaced with albumin and supplements such as insulin, selenium and transferrin.

Buffering of culture media for mammalian cells is usually provided by sodium bicarbonate. This disindirectly with cultured mammalian cells must be chemically clean as well as sterile. Immersion in nitric or hydrochloric acid is an effective method of cleaning glassware; most detergents cannot be used since they tend to persist on glass surfaces. However cleaned, all apparatus must be exhaustively rinsed with high-quality distilled water. For sterilization of apparatus dry heat can be used but large assemblies are usually steam-sterilized.

Growth medium is prepared in vessels constructed from glass or high-grade stainless steel. High-quality water of the desired volume is mixed with the chemical ingredients of the medium, serum is added and the pH value adjusted if necessary. Because some of the components of the medium will be heat-labile, sterilization is done by filtration.

MODES OF CELL GROWTH

Under optimal conditions primary epithelial cells form layers one cell thick while the lawns formed by fibroblasts are usually 2–3 cells thick. Such cells generally only grow when attached to surfaces. The same principles apply for cell strains and cell lines derived from normal cells. Only rarely can cell lines be derived that multiply well in suspension and these usually arise from transformed cell lines. Cultures of cells which grow well in suspension produce many more cells than those whose growth is restricted to surfaces. Transformed cells and cells of some lines can grow and form colonies on semisolid media, e.g.

agarose, and this is particularly useful when cloning cells, as in hybridoma selection (see p. 130). These different modes of growth are reflected in the methods used to culture cells in the laboratories.

SMALL-SCALE CELL CULTURE

Primary cell cultures and cell strains are grown either in plastic Petri dishes (tissue culture grade) or in plastic T-flasks (Fig. 8.2). The latter are a modification of the Carrel flask introduced by Alexis Carrel in 1923. These T-flasks are available in a variety of sizes with surface areas ranging from 25 to 175 cm^2. In a typical experiment a 75 cm^2 T-flask containing 15–20 ml of medium would be seeded with 10^6 cells and incubated at 37 °C. After 2–3 days the bottom surface of the flask will be covered with a confluent monolayer of cells. The cells can readily be detached by trypsin or EDTA treatment.

The reason why a high cell density is required in the inoculum is related to the fact that for recently isolated cells the medium must contain adequate concentrations of unknown factors produced by the cells themselves. This is not so true for transformed cells, which can withstand far greater dilution. The efficiency of plating of recently isolated cells, e.g. primary cultures and cell strains, can be greatly increased if they are mixed with a feeder layer of similar cells made incapable of multiplication by X-irradiation; these cells are still metabolically active and supply factors that enable the non-irradiated cells to survive and

Fig. 8.2 T-flask (left) used for the culture of anchorage-dependent cells and spinner bottle (right) used for suspension culture.

multiply. The efficiency of plating can also be increased by Earle's technique of introducing individual cells into very small volumes of medium, e.g. in microtitre dishes. The cell-produced factors thus reach an adequate concentration without the need for a feeder layer.

Although transformed cells also can be grown in T-flasks they can be grown more efficiently and effectively in suspension in spinner bottles. These spinner bottles (Fig. 8.2), which are simply bottles containing a captive magnetic stirrer bar, are available in a variety of sizes from 25 ml to 9 l. Bottles are inoculated with cells at a density of at least 0.25×10^6 cells/ml and purged with an air/CO_2 mixture. The contents are stirred magnetically but sufficiently gently for shear damage to the cells not to occur; a satisfactory speed is achieved when a liquid vortex is just beginning to appear on the stirrer shaft.

LARGE-SCALE CELL CULTURE [1, 2, 3, 4, 5, 6, 7]

The large-scale production of mammalian cells which grow in suspension culture is not very difficult. Standard laboratory or production fermenters can be used since these are designed to run aseptically for long periods and have facilities for gassing and pH control. However, the stirrer motor may have to be geared down to ensure that a sufficiently low speed of stirring can be obtained to prevent undesirable shear effects on the cells. An alternative method of culturing the cells is to use an airlift fermenter (p. 70); the sparged gas agitates the cells sufficiently to promote metabolite interchange without causing damage to the cells.

The growth of anchorage-dependent cells on a large scale is much more difficult, for a fermenter does not provide the requisite high surface area to volume ratio. A number of different methods have been developed and the choice of method depends partly on the use to which the cells will be put and partly on the capital which can be spent. The simplest system is the use of roller bottles, which have a capacity of 500 ml to 50 l. These are filled with about one-twentieth of their volume of medium, inoculated with cells and placed on their sides on slowly revolving rollers. As the bottles revolve the growing monolayer is alternately immersed in the liquid medium and exposed to the air in the bottle. Thus the cells are automatically washed with nutrients and exposed to oxygen and the movement of the liquid medium ensures uniform distribution of its components. As can be readily envisaged,

large-scale roller bottle operation involves simple technology but is labour-intensive. In addition, the bottles are constructed from high-quality glass and thus are expensive but they have only a finite life; eventually cells fail to adhere to the inner surface.

One way of increasing the surface area to volume ratio in a fermenter is to fill it with a series of closely spaced plates. Various designs have been described but they all suffer from the same disadvantages: high construction costs, poor mixing and difficulty in encouraging cells to stick to the plate surfaces. A much better way of utilizing conventional fermenters is to make use of microcarrier beads. First described in 1967 by Anton van Wezel many variations of the microcarrier method have been exploited on an industrial scale. The beads are composed of dextran or a synthetic polymer and are covered with a large number of functional groups, e.g. DEAE groups. Since the beads are only $50-200\,\mu m$ in diameter and have a density slightly greater than that of water they are readily maintained in suspension by gentle agitation. Anchorage-dependent cells inoculated into the medium bind to the beads and multiply. The advantage of this system over conventional roller bottles can be seen by reference to Table 8.6. Porous microcarriers are now available. These are preformed particles with a porous internal structure into which animal cells can grow. They can be maintained in suspension in culture medium. The advantages of porous microcarriers are that they can support a higher cell density per unit of microcarrier bed volume than solid microcarriers and that the matrix protects cells against shear in agitated cultures.

A completely different method of growing anchorage-dependent cells involves the use of hollow-fibre tech-

Table 8.6 Comparison of different culture methods for anchorage-dependent cells.

Roller bottle	Surface area	$4.9 \times 10^2\,cm^2$/bottle
Microcarriers	Surface area	$6 \times 10^3\,cm^2$/g dry wt
	Normal density	5 g dry wt/l
	Surface area/l	$3 \times 10^4\,cm^2$
	Roller bottle equivalents	60
Hollow-fibre apparatus	Size	$40 \times 40 \times 4.5\,cm$
	Fibre surface area	$9.3 \times 10^3\,cm^2$
	Roller bottle equivalents	19

nology. The cells are grown on the outer surface of the fibres, which are immersed in the growth medium, while air and carbon dioxide are passed through the lumen of the fibres. In practice, flat bed structures are used with the fibres no more than three to six layers deep. This arrangement ensures that when medium is pumped *through* the bed the path of perfusion is short. Thus there is hardly any gradient from a high to a low nutrient level and from a low to a high waste product level. Once the cells have encrusted the fibres it is possible to switch from a growth medium to a maintenance medium. The latter medium contains no serum, and hence it is much cheaper, but it still permits the secretion of macromolecules for prolonged periods.

PROBLEMS ASSOCIATED WITH LARGE SCALE CELL CULTURE

From the foregoing section it should be clear that the large-scale culture of animal cells can be achieved but that it is not without its problems. Some of these problems are summarized in Table 8.7.

GENERAL GUIDELINES FOR THE USE OF ANIMAL CELL CULTURES

Listed below are some specific recommendations for the handling of animal cell cultures, which are not covered elsewhere in this chapter.

1 The cell culture facility should be designed specifically for the purpose and should be adequately equipped to permit proper aseptic laboratory procedures.
2 Whenever possible, cell lines should be obtained only from reputable repositories, who should be able to provide details of the history of the culture and the quality control checks which have been performed on it.
3 When new cell lines arrive in the laboratory they should be held in quarantine, preferably in a separate area, until sterility and other tests have been concluded. Mycoplasma assays are particularly important and should include positive and negative controls.
4 Once a new culture has been shown to be free of adventitious agents, a series of stocks should be frozen down.
5 Medium components, especially serum, should be pretested for growth promotion with appropriate cell cultures.
6 All contaminated media and infected cultures should be disposed of promptly and properly.
7 Cell lines should be checked routinely to ensure that they exhibit no changes in karyotype, phenotypic expression or growth pattern.

Further reading

GENERAL REFERENCES

Feder J. & Tolbert W.R. (1983) The large-scale cultivation of mammalian cells. *Scientific American* **248**, 24−31.
Harakas N.K. (1984) Industrial mammalian cell culture: physiology−technology−products. *Annual Report of Fermentation Processes* **7**, 159−211.
Jakoby W.B. & Pastan I.H. (eds) (1979) *Methods in Enzymology*, vol. 58, *Cell Culture*, Academic Press, New York.
Thilly W.G. (ed.) (1986) *Mammalian Cell Technology*. Butterworths, London.

CITED REFERENCES

1 Clark J.M. & Hirtentein M.D. (1981) Optimizing culture conditions for the production of animal cells in microcarrier culture. *Annals of the New York Academy of Sciences* **369**, 33−46.
2 Cahn F. (1990) Biomaterials aspects of porous microcarriers for animal cell culture. *Trends in Biotechnology* **8**, 131−6.
3 Kearns M.J. (1990) Integrated design for mammalian cell culture. *Bio/technology* **8**, 409−13.
4 Scheirer W. (1987) Laboratory management of animal cell culture processes. *Trends in Biotechnology* **5**, 261−5.
5 Familletti P.C. & Fredericks J.E. (1988) Techniques for mammalian cell immobilization. *Bio/technology* **6**, 41−4.
6 Hopkinson J. (1985) Hollow fibre cell culture systems for economical cell-product manufacturing. *Bio/technology* **3**, 225−30.
7 Ku K., Kuo M.J., Delente J., Wildi B.S. & Feder J. (1981) Development of a hollow fibre system for large-scale culture of mammalian cells. *Biotechnology and Bioengineering* **23**, 79−95.

Table 8.7 Problems associated with large-scale mammalian cell culture.

1 Cost and availability of good-quality mammalian serum
2 Potential contamination of bovine serum with foot and mouth disease virus (FMDV). This is not a problem in the US, which is FMDV-free
3 Potential contamination of cells with mycoplasmas or even animal viruses of human origin. Such infections are not always easy to detect
4 Large-scale use of antibiotics can be costly and they can contaminate the final product. In the case of penicillin residues this would not be acceptable
5 Require large amounts of high-quality distilled water and steam, which should be pyrogen-free. This is costly
6 May require special equipment or culture methods for anchorage-dependent cells

9 Applications of Animal Cell Culture

Scientists studying the biochemistry and biophysics of cell growth and division have made extensive use of cultured animal cells, as have those engaged in research on animal viruses. However, the value of cultured animal cells goes far beyond this, for they produce a wide range of biological products of commercial interest including immunoregulators, antibodies, polypeptide growth factors, enzymes and hormones. Already they are used in the manufacture of virus vaccines, tissue plasminogen activator (an enzyme which facilitates the destruction of blood clots), interferon-α (for the treatment of cancer), monoclonal antibodies and tumour-specific antigens (for inclusion in diagnostic kits). Consideration is also being given to the possibility of growing fibroblasts in culture from, for example, burns patients. These cultured cells would be used as a reconstituent skin to facilitate wound healing. Another potential application of cultured cells is their use in evaluating new drugs and toxic chemicals. When used in concert with conventional animal models they could reduce much of the time, cost and effort as well as the total number of animals. Even the cells themselves can have commercial value. For example, cultured fibroblasts from burns patients have been used as a reconstituted skin to facilitate wound healing.

IN VITRO TOXICITY TESTING [1]

The use of animals in evaluating chemical safety is costly, time-consuming and increasingly criticized by animal welfare groups. As a consequence, alternative test methods are being developed which make use of cultured cells. For example, genotoxicity can be assessed by using incorporation of radioactive nucleotides as a measure of DNA repair. Not only must chemicals used on large scale, or administered as pharmaceuticals, be free of cyto- or genotoxicity but so must their metabolites. Rather than administer such compounds to test animals, they can be added to cultures of animal or human hepatocytes (liver cells).

PRODUCTION OF VIRAL VACCINES

The oldest commercial application of animal cell cultures is for the production of viral vaccines and it could be argued that globally, with the possible exception of antibiotics, they have provided greater benefit than any other pharmaceutical product. The deployment of a single vaccine, that against smallpox, has been of immense benefit to mankind and hopefully has eliminated the virus for evermore. Because viruses are obligate intracellular parasites the earliest viral vaccines, those against smallpox and rabies, were made in intact animals such as calves, sheep and rabbits. Today the only intact host used in advanced production techniques is the developing chick embryo for the manufacture of influenza and yellow fever vaccines. All other vaccines are prepared by the growth of virus in cell culture. Table 9.1 lists the major human and veterinary virus vaccines and the cell cultures used in their production.

Viral vaccines are of two types. Those which are prepared from live, avirulent (attenuated) organisms often give lifelong protection but carry with them the attendant risk of contamination with other viruses derived from the cell culture. Clearly, rigorous microbiological and other quality control procedures are essential. Vaccines prepared from killed (inactivated) viruses are less satisfactory since they are not as immunogenic as live vaccines, presumably because the virus does not multiply in the animal host, and have to be injected repeatedly according to carefully balanced immunization schedules. The only inactivated vaccines for human use are those against rabies and, in some instances, polio. The usual inactivating agents are dilute formalin or β-propiolactone.

In theory, the production of viral vaccines is relatively simple. The cell cultures principally used for virus vaccine production are prepared from monkey kidney or chick embryos. More recently there has been an increasing use of human diploid cells. Since these cells grow as monolayers all that is required in vaccine

Table 9.1 Some viral vaccines currently available for human and veterinary use.

	Human vaccines		Veterinary vaccines	
Virus	Cell used for culture	Virus	Cell used for culture	
Measles	Chick embryo fibroblasts	Canine distemper	Chick embryo fibroblasts or dog kidney cells	
Polio (inactivated)	Monkey kidney cells	Canine hepatitis	Dog kidney cells	
Polio (live)	Monkey kidney or human diploid cells	Foot and mouth disease	Bovine kidney cells	
Rabies	Human diploid cells	Rabies	Duck embryo or chick embryo	
Rubella	Rabbit kidney, duck embryo or human diploid cells	Feline panleucopenia	Cat kidney cells	
		Marek's disease	Chick embryo cells	

preparation is to infect them with the appropriate virus and then harvest the culture fluid after virus multiplication has occurred. The culture fluid containing the virus is clarified by filtration. In the case of killed-virus vaccines a concentration step is usually employed after the inactivation step. When processing is complete the virus suspension is blended with stabilizers to prevent loss of potency, a problem with live attenuated vaccines, and stored at low temperature. A typical production process is shown in Fig. 9.1.

In practice, the production of a virus vaccine is far more complex, not just because of the labour involved in large-scale animal culture but because of the rigorous quality control and safety procedures which are required; for example, all tissue culture substrates are rigorously examined to exclude contamination with infectious agents from the source animal or, in the case of human diploid cells, to exclude abnormal cellular characteristics. Monkey tissues are tested exhaustively for extraneous agents, e.g. herpes virus B, simian virus 40, tubercle bacilli and mycoplasmas. Human diploid cells are subjected to detailed karyological examination to exclude cultures with features resembling transformed cell lines or malignant tissues.

Safety of the final product is another important consideration. With killed-virus vaccines the potential hazards are those due to incomplete virus inactivation —a possible cause of recent outbreaks of foot and mouth disease in Europe! The tests used to detect live virus consist of the inoculation of susceptible tissue culture and of susceptible animals. The cultures are examined for cytopathic effects and animals for symp-

toms of disease and histological evidence of infection at autopsy. With attenuated viral vaccines the potential hazards are those associated with reversion of the virus to a degree of virulence capable of causing disease in vaccinees. This possibility is minimized by the use of stable virus seed stocks but virulence checks on the final product are necessary; for example, in the production of attenuated poliomyelitis vaccine the neurovirulence of each batch following intraspinal inoculation of monkeys is compared with that of a control vaccine.

Many of the problems associated with the production of viral vaccines, particularly the safety aspects, may disappear now that recombinant DNA technology can be used to enable virus subunits to be produced in bacteria or yeasts. This has been discussed already in Chapter 6. Another advantage of recombinant DNA technology is that it may permit the production of vaccines against those viruses which cannot be grown in cell culture to a titre high enough to provide sufficient antigen for effective vaccination. An example based on hepatitis B virus was presented in Chapter 6 (see p. 94).

PRODUCTION OF HIGH-VALUE THERAPEUTICS [2, 3]

There are many human proteins which have long been believed or known to have therapeutic potential but which are in short supply; for example, human growth hormone can be obtained from the pituitary gland of cadavers and clotting factors, VIII and IX from the

APPLICATIONS OF ANIMAL CELL CULTURE   119

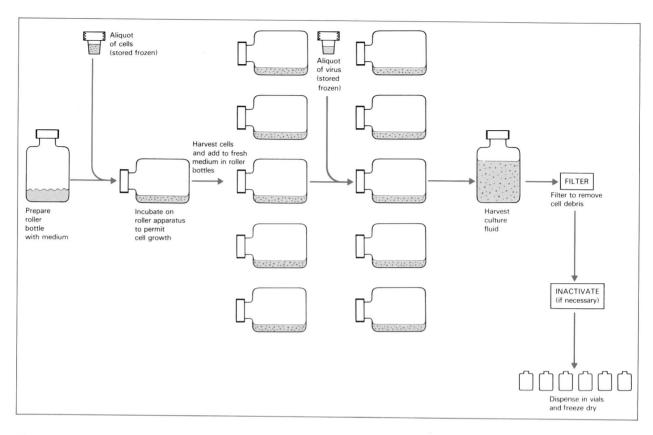

Fig. 9.1 Production of killed-virus vaccines.

Table 9.2 Proteins overproduced by some mammalian cell lines.

Product	Cell line name	Source
Acetylcholinesterase	BP4 1AS	Murine neuroblastoma
Interferon-α	Namalwa	Human blood
Interferon-β	Flow 7000	Human embryonic foreskin
Interleukin	EL4 CL14	Murine blood
Plasminogen activator	GPK	Guinea pig hepatocyte
	BEB	Human breast
Urokinase	HT 1080	Human fibrosarcoma

plasma of blood donors. The supply limitations of these substances is obvious and there are other complications, e.g. the potential for acquiring AIDS from contaminated blood products. A more satisfactory alternative would be to grow, in large-scale culture, cells derived from the tissue which normally synthesizes the desired protein. Some examples are given in Table 9.2. The disadvantage with this method is that, generally speaking, such cultured cells produce only low levels of the therapeutic protein: interferon-β is a good example. Using human embryonic lung fibroblasts growing in 20 l roller bottles, each containing 750 ml of medium, only 5×10^6 units (0.02 mg) of interferon-β are produced per litre of medium *after* induction of interferon synthesis with polyI:C. Until the advent of suitable monoclonal antibodies the purification of this interferon, free of serum proteins from the growth medium, was very difficult.

One solution to the problem of low yields is to produce the protein in bacteria by using recombinant DNA technology as outlined in Chapter 6. The problem with this approach is that many mammalian proteins undergo post-translational modifications by processes such as glycosylation which are not mimicked in bacterial systems. Although many of the non-glycosylated derivatives of these proteins retain activity, e.g. interferon-γ, urokinase and interleukin-2, their half-life *in vitro* may be reduced significantly and long-term therapy in humans may result in undesirable immunological reactions. With some proteins, such as human blood clotting factor IX, the situation is more complex: not only is glycosylation essential for activity but so too is the vitamin K-dependent γ-carboxylation of glutamic acid residues. Clearly, at least for the foreseeable future, the only prospect of manufacturing such molecules is to do so in mammalian cells. However, this will not solve all the above problems. When human protein C was produced by three different cell lines, the levels of γ-carboxylation and glycosylation varied significantly, as did the antithrombotic activity. The method of cell culture can also affect the glycosylation pattern significantly as well as the *in vivo* half-life. When IgM was produced in serum-supplemented perfusion culture it had an *in vivo* half-life of 18 minutes but from serum-free culture it was only 3 minutes.

If mammalian cell culture is the only method for obtaining certain therapeutic proteins, then cells which are overproducers are required. It is well known that in some pathological conditions certain cells in the body become neoplastic and overproduce a particular protein; Table 9.2 lists some of the proteins overproduced by transformed cells. However, transformed cells that will produce many of the mammalian proteins currently in demand are not available. The solution is to clone the relevant genes in mammalian cells rather than microbial cells.

CLONING IN ANIMAL CELLS: BASIC PRINCIPLES

Both bacteriophage and plasmid vectors are available for gene manipulation in bacteria but because of ease of use the latter are favoured. However, plasmids do not occur naturally in animal cells and so only virus vectors are available. As for any vector, the virus genome must be easily manipulated, it must contain convenient sites for the restriction enzymes used in cloning and the location of those sites with respect to vector control sequences must be known. This latter point is particularly important since the gene sequences required for virus replication are far more complex than those required for plasmid replication. With virus vectors there can be an additional compli-

cation: the need to package the recombinant genome in viral coat protein. This can lead to size constraints on the amount of foreign DNA which a vector can carry. For this reason considerable emphasis is being placed on the development of vectors which can replicate like plasmids and thus at no stage need to be encapsidated in virus particles.

Whatever kind of vector is used there will always be a need at some stage to introduce unpackaged DNA into animal cells in culture, a procedure known as *transfection*. The most commonly used procedure is dependent upon the fact that freshly prepared co-precipitates of DNA and calcium phosphate are phagocytosed by cells. The method is not particularly efficient since only a low proportion of cells take up the exogenous DNA; at best only 1–2% of cells are infected but the efficiency can be as low as 1 cell in 10^6. When virions are produced the efficiency of transfection is not a limiting factor as the recombinant virions can subsequently be used to guarantee efficient infection.

It is not essential to have vectors capable of replication in order to stably introduce DNA into mammalian cells. Fragments of DNA can be phagocytosed by means of the calcium phosphate coprecipitation method and in at least a few of the transfected cells will become stably integrated at random in the genome. Identification of these rare cells carrying exogenous DNA is difficult unless the DNA also carries a gene for a selectable marker. A range of selectable genes is available (Table 9.3) for ligating to test genes prior to transfection. This technique has been used to obtain expression of hepatitis B surface antigen, herpes simplex type I glycoprotein D and human interferon-γ in Chinese hamster ovary cells. However, even though multiple copies of the transfected gene are present, expression levels are low and this technique is better suited to gene therapy (see Chapter 11).

Table 9.3 Some markers whose presence on exogenous DNA facilitates the selection of transfected cells.

Selectable marker	Basis of selection
Thymidine kinase (from herpes simplex virus)	Permits growth of cells lacking thymidine kinase
Bacterial neomycin phosphotransferase	Confers resistance to aminoglycoside antibiotics, e.g. G418
Bacterial dihydrofolate reductase	Confers resistance to methotrexate

VIRUS VECTORS FOR USE WITH MAMMALIAN CELLS [4, 5, 6]

A number of different virus vectors are available for cloning in mammalian cells. Each vector system has its advantages and disadvantages and, as yet, no one type of vector has emerged as clear favourite. To some extent this is a reflection of the different host ranges of the different viruses but personal preference also plays a part. The best example of this comes from work on the expression of clotting factor IX where three different vector systems have been used successfully. Only a brief description of each system is presented below and for a more detailed analysis the reader should begin by consulting Old and Primrose (1990). It should be borne in mind that the development of animal virus vectors is progressing at a phenomenal rate.

SV40 vectors. Simian virus 40 (SV40) was the first eukaryote virus to be harnessed as a vector simply because it was the first such virus for which a complete nucleotide sequence was available. The genome of SV40 contains very little non-essential DNA so it is necessary to insert the foreign gene in place of essential viral genes in order not to exceed the size limits imposed by the viral protein coat. Consequently the recombinant genome can be propagated only in the presence of a helper virus. In the original SV40 vectors the gene that is replaced is one which is required late in infection. The procedure for using such a vector is to perform a mixed infection between the vector carrying the foreign gene insert and a helper virus carrying a temperature-sensitive mutation in a gene required early in infection. Following infection at the non-permissive temperature virus growth can occur only by complementation between the helper virus, which provides late functions, and the recombinant, which provides early functions.

An alternative SV40 vector system is one in which the foreign gene insert replaces a gene required early in infection. The advantage of this system is that, although these recombinant viruses are unable to replicate in most monkey cell lines, they can do so in a special line of cells known as COS. Since helper virus is not required this means that pure recombinant virus stock can be produced easily. However, expression of foreign genes from the early promoter is much lower than from the later promoter.

A whole series of vectors has been constructed which comprise part of the *E. coli* vector plasmid pBR322 and part of the SV40 genome and which permit the exogenous gene to be placed under the control of an SV40 promoter. These vectors are grown and manipulated using *E. coli* as the host and are then transferred back into monkey cells. A similar series of vectors based on a related virus, polyoma, is available for use with cultured mouse cells.

Adenovirus vectors. The major disadvantage with SV40- and polyoma-based vectors is the small size of the genome. Since packaging constraints are strict, only small gene inserts can be made. Adenoviruses are much larger than SV40 and can accommodate much more extraneous DNA. However, the larger size means that the adenovirus genome is intrinsically more difficult to manipulate since it contains multiple sites for restriction endonucleases. Using a variety of techniques, vectors are being constructed in which many of these sites have been eliminated. In one set of vectors exogenous DNA is placed under the control of a transcriptional unit which is functional early in the life-cycle of the virus. Although the exogenous gene replaces a viral gene function, no helper virus is required if the recombinants are grown in a special strain of human embryonic kidney cells.

Vaccinia. Vectors based on vaccinia virus are used in a slightly different way. Exogenous genes are first cloned adjacent to a vaccinia promoter present in the middle of a vaccinia-derived thymidine kinase gene which is carried on an *E. coli* plasmid (Fig. 9.2). After manipulations are complete in *E. coli* the recombinant plasmid and wild-type vaccinia virus are cotransfected into a monkey cell line. Recombination occurs between the plasmid-borne thymidine kinase gene fragment and the viral thymidine kinase gene. Recombinant virions fail to synthesize thymidine kinase and can be selected by virtue of their resistance to 5-bromodeoxyuridine.

BPV vectors. Vectors based on bovine papillomavirus (BPV) are unique in that not only are they propagated as multicopy plasmids but they also cause neoplastic transformation of the host cell, i.e. the cell becomes cancerous. The earliest vectors comprise a large subgenomic transforming fragment of the BPV genome linked to an *E. coli* plasmid. Selection of cells carrying the vector is easy: being neoplastic they outgrow the normal cells, which only form a monolayer in culture. BPV-derived vectors show great promise because per-

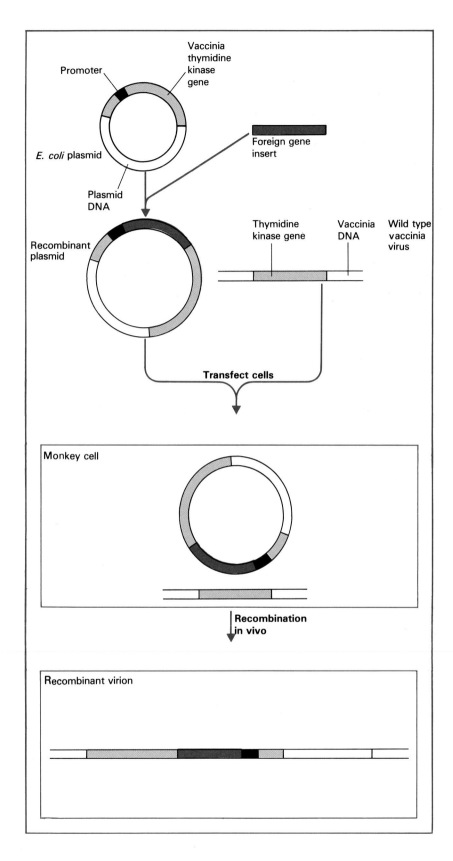

Fig. 9.2 The insertion of a foreign gene into the vaccinia genome by recombination *in vivo* (see text for details).

manent cell lines are obtained which carry the foreign DNA insert at high copy number (up to 200 per cell). An additional advantage is that there is no size limitation on the DNA insert.

Baculovirus vectors. Baculoviruses infect insects. They have large double-stranded circular DNA genomes within a rod-shaped protein coat. A property of those baculoviruses which have been exploited as vectors is that during normal infection they produce nuclear inclusion bodies. These consist of virus particles embedded in a protein matrix of a virus-encoded protein called polyhedrin. Large amounts of virus and polyhedrin are produced. Transcription of the polyhedrin gene is driven by an extremely active promoter, which can be used to drive expression of foreign genes. This is particularly attractive because the polyhedrin gene product is not essential for viral replication. Construction of expression vectors has consisted of inserting a foreign coding sequence just downstream of the polyhedrin promoter. This cannot be achieved directly because the large size (about 130 kb) of the viral DNA precludes simple *in vitro* manipulation. Thus the strategy for inserting the foreign DNA into the virus is similar to that described above for vaccinia virus.

Baculovirus vectors are available for use either with cultured insect cells or silkworms. The silkworm host has advantages as a production system because silkworms can be cultured easily and at low cost as in the long-established silk industry.

Retrovirus vectors. Retroviruses have single-stranded RNA genomes and, at first sight, would appear to be unpromising candidates as vectors but the peculiarities of the retrovirus life-cycle mean that these viruses provide perhaps the most promising vector system of all. During the process of reverse transcription, a double-stranded proviral DNA molecule is generated which has all the regulatory functions for viral transcription and translation (see Box. 9.1). Thus, any of the normal proviral genes can be replaced with foreign DNA provided the recombinant is propagated in the presence of a helper virus. Retroviruses have a number of features which make them particularly useful as gene transfer vectors and these are summarized in Table 9.4.

PRODUCTION OF THERAPEUTICS AND VACCINES IN RECOMBINANT MAMMALIAN CELLS [6, 7, 8, 9, 10, 11]

A whole range of proteins of potential value to the pharmaceutical industry has now been produced in cultured, recombinant mammalian cells. Some of them are listed in Table 9.5 and several are approved for clinical use. Certain specialized antibodies have also been produced in a similar fashion and these are described in the next chapter.

All of these proteins are secreted as glycoproteins but for at least three of them (tPA, interferon-β and erythropoietin) the pattern of glycosylation differs from that seen in the natural molecule. However, since the glycosylation pattern differs for the same protein made in different cell types it is questionable as to the definition of the normal glycoform.

Vaccinia vectors have been extensively used to develop novel vaccines. Some proteins expressed by recombinant vaccinia viruses are listed in Table 9.6. The vaccinia virus recombinant expressing the influenza haemagglutinin gene was able to induce resistance to influenza virus infection in hamsters. More interesting is the work done with the rabies antigen recombinant. This was administered to the wild population of foxes in north-eastern France by providing 'bait' consisting of chicken heads spiked with the virus. Subsequently a substantial proportion of the wild foxes were shown to have acquired immunity to rabies. The same virus has been used to eliminate rabies in wild animal populations in other European countries and is being trialled in the USA.

Whether recombinant vaccinia viruses will ever be used for human vaccination remains to be seen. Interestingly in this context, an unplanned trial of

Table 9.4 Properties of retroviruses which are of value in gene cloning.

1 The normal replication process involves the stable insertion of a DNA copy of their genome into the host genome
2 They can infect and transmit their genetic information into a high proportion of recipient cells
3 Retroviruses have a broad host range which can be easily modified, making it possible to infect cells from a variety of species and cell types
4 The integrated provirus exists as a stable, predictable structure within the host genome
5 Infection of mammalian cells with retroviruses does not result in cell death. Rather, infected cells continue to grow and divide while producing large numbers of virus particles
6 A large amount of foreign gene sequence can be packaged within a virus particle
7 Viral gene expression is driven by a strong promoter

Box 9·1 The biology of retroviruses

Retroviruses differ from all other viruses of animals in that their genome is diploid, consisting of two identical molecules of single-stranded RNA held together at their 5'-ends by hydrogen bonding. The virus particles also contain various species of low-molecular-weight RNA molecules including tRNA. The function of most of these smaller RNA molecules is not known but the tRNA serves as a primer for DNA replication (see Box. 3.1, p. 29).

Three viral genes are essential for replication: *gag*, which encodes an internal structural protein of the virus; *pol*, which encodes an RNA-dependent DNA polymerase; and *env*, which encodes glycoproteins of the virus envelope. Viruses that carry all three of these genes are replication-competent and are termed non-defective. Many retroviruses exist which lack some or all of these essential genes and such *defective* viruses depend on non-defective helper viruses for replication. Many of the retroviruses contain a gene, *onc* or *src*, which is homologous to a normal cellular gene and whose overexpression can result in cellular transformation. Finally, most retroviruses have a *c* region which carries an efficient promoter that causes viral products to be made in large quantities. Diagram 1 shows the gene order in a typical non-defective avian retrovirus.

The replication cycle of retroviruses can be divided into three parts (Diagram 2): first, the conversion of single-stranded viral RNA to double-stranded (proviral) DNA; second, integration into the chromosome; and, third, transcription and translation of the integrated provirus. In addition to being translated, the RNA which is transcribed may be packaged into virus particles.

The RNA-dependent DNA polymerase, or reverse transcriptase, encoded by the *pol* gene, has two enzymatic activities: a DNA polymerase capable of copying RNA or DNA templates and a ribonuclease (RNase H) that specifically removes the RNA from RNA:DNA hybrids. This reverse transcriptase and another enzyme (integrase), which catalyses insertion of the provirus into the host genome, are packaged into the virion along with the viral RNA.

Diagram 2 The replication of the retrovirus genome.

Diagram 1 The structure of the retrovirus genome.

Table 9.5 Some mammalian proteins overproduced in cells in culture as a result of *in vitro* gene manipulation.

Protein	Size	Use	Comments
Tissue plasminogen activator	527 amino acids	Thrombosis	Enzyme overproduced in *E.coli* or yeast not fully functional
Interleukin-2	133 amino acids	Cancer therapy	
Tumour necrosis factor	157 amino acids	Cancer therapy	
Factor VIII	2332 amino acids	Haemophilia	Normally obtained from plasma of blood donors but now there is concern over potential contamination with AIDS virus
Factor IX	415 amino acids	Christmas disease	Must be made in mammalian cells as glycosylation and conversion of first 12 glutamate residues to pyroglutamate essential for activity
Erythropoietin	166 amino acids	Anaemia	Without glycosylation protein is cleared very quickly from plasma
Interferon-β	164 amino acids	Cancer therapy	Many other clinical indications
Granulocyte–macrophage colony-stimulating factor		Stimulates production of white blood cells after cancer chemotherapy	

Table 9.6 Viral genes expressed in vaccinia recombinants.

Hepatitis B virus surface antigen
Herpes simplex virus glycoprotein D
Rabies virus glycoprotein G
Vesicular stomatitis virus (VSV) nucleoprotein
Epstein–Barr virus glycoprotein
Influenza virus haemagglutinin
Malaria sporozoite antigen
Human respiratory syncytial virus glycoprotein G
Sindbis virus structural proteins
HTLV-III (AIDS virus) envelope gene

recombinant vaccinia in man has been reported. A researcher accidentally inoculated a small cut on his finger with a drop of recombinant vaccinia which expresses the vesicular stomatitis virus (VSV) nucleoprotein. VSV causes a highly contagious vesicular disease of cattle, pigs and horses and flu-like symptoms in man. It was evident that the researcher had become infected with the recombinant but he did not experience malaise. The mildness of the infection was notable because a non-vaccine strain of the virus was involved. This could be related to the fact that the researcher had been vaccinated some 30 years previously, or to attenuation of the virus by the insertional inactivation of the viral *tk* gene. Importantly, serum from the researcher showed a high antibody titre to the VSV protein. This event showed that despite previous vaccination, a new immune response was elicited to non-vaccinia protein in man.

Further reading

GENERAL REFERENCES

Cartwright T. (1987) Isolation and purification of products from animal cells. *Trends in Biotechnology* **5**, 25–30.
Chanock R.M. & Lernes R.A. (1984) *Modern Approaches to Vaccines: Molecular and Chemical Basis of Virus Virulence and Immunogenicity.* Cold Spring Harbor Laboratories, Cold Spring Harbor.
Old R.W. & Primrose S.B. (1990) *Principles of Gene Manipulation: an Introduction to Genetic Engineering*, 4th edn. Blackwell Scientific Publications, Oxford.
Pavlakis G.N. (1990) Gene expression using DNA viral vectors. *Current Opinion in Biotechnology* **1**, 48–54.
Speir R.E. & Griffiths J.B. (1985) *Animal Cell Biotechnology*, vols 1 and 2. Academic Press, London.

CITED REFERENCES

1 Goldberg A.M. & Frazier J.M. (1989) Alternatives to animals in toxicity testing. *Scientific American* **261**, 16–22.
2 Yan S.C.B., Razzano P., Chao Y.B., Walls J.D., Berg D.T., McClure D.B. and Grinnell B.W. (1990) Characterization and novel purification of recombinant human protein C from three mammalian cell lines. *Biotechnology* **8**, 655–61.
3 Van Brunt J. (1990) The importance of glycoform heterogeneity. *Biotechnology* **8**, 995.
4 Graham F.L. (1990) Adenoviruses as expression vectors and recombinant vaccines. *Trends in Biotechnology* **8**, 85–7.
5 Miner J.N. & Hruby D.E. (1990) Vaccinia virus: a versatile tool for molecular biologists. *Trends in Biotechnology* **8**, 20–5.

6 Hruby D.E. (1990) Vaccinia virus vectors: new strategies for producing recombinant vaccines. *Clinical Microbiology Reviews* **3**, 153–70.

7 Bolognesi D.P. (1990) Approaches to HIV vaccine design *Trends in Biotechnology* **8**, 40–5.

8 Parekh R.B., Dwek R.A., Edge C.J. & Rademacher T.W. (1989) N-glycosylation and the production of recombinant glycoproteins. *Trends in Biotechnology* **7**, 117–22.

9 Blancou J., Kieny M.P., Lathe R., Lecocq J.P., Pastoret P.P., Soulebot J.P. & Desmettre P. (1986) Oral vaccination of the fox against rabies using a live recombinant vaccinia virus. *Nature* **322**, 373–5.

10 Jones L., Ristow S., Yilma T. & Moss B. (1986) Accidental human vaccination with vaccinia virus expressing nucleoprotein gene. *Nature* **319**, 543.

11 Okamoto M., Nakayama C., Nakai M. & Yanagi H. (1990) Amplification and high-level expression of a cDNA for human granulocyte–macrophage colony-stimulating factor in human lymphoblastoid Namalwa cells. *Bio/technology* **8**, 550–3.

10 Monoclonal Antibodies

When a foreign macromolecule (*antigen*) is introduced into the circulatory system of a higher vertebrate it stimulates specific white blood cells, the lymphocytes, to produce antibodies that combine specifically with the macromolecule to facilitate its destruction within, or removal from, the body. The antibodies which are synthesized can be found in the globulin fraction of the proteins that circulate in the blood and hence are called *immunoglobulins*. All immunoglobulin molecules have a similar basic structure consisting of two heavy and two light chains held together by disulphide bonds (Fig. 10.1). Regions of each chain are associated with particular functions. The *variable (V) regions* are responsible for antigen binding. Within the variable regions are relatively constant framework sequences which seem to act as a scaffold for the hypervariable regions, also known as *complementarity-determining regions* (CDR), which are largely responsible for antibody specificity. Each different antibody has a different amino acid sequence and spatial arrangement and it is conceivable that every possible shape presented by an antigen can be accommodated by some antibody produced by the immune system. The heavy chain constant regions interact with receptors on

Fig. 10.1 Schematic representation of an immunoglobulin molecule.

various cells involved in the immune response and interact with other proteins. These functions are associated with particular domains (C_H1, C_H2, C_H3) of the constant region. When an antibody molecule is treated with papain, the heavy chain is cleaved once to produce three fragments, two antigen-binding fragments (Fab) and a constant-region fragment (Fc).

When the immune system of an animal encounters a new antigen it does not respond to the entire surface of the antigen but to specific antigenic determinants (*epitopes*) located on it. Thus a protein antigen may possess several epitopes and would induce the formation of several different antibodies; an object the size of a bacterium would possess many different epitopes and induce the synthesis of a correspondingly greater number of antibodies. Following induction of antibody formation the animal can be bled and the serum fraction obtained. However, this serum will contain all the different antibodies produced in response to the antigen. Furthermore these antigen-specific antibodies will be diluted with all the other antibodies present in the serum as a result of the animal's previous encounters with other antigens. Such polyclonal antisera have been widely used by biologists but, as will be seen later in this chapter, there are many commercial applications of antibodies where it is essential to have only a single antibody species. These *monoclonal antibodies* can be produced by hybridoma technology.

MONOCLONAL ANTIBODIES FROM HYBRIDOMA TUMOURS [1]

Antibodies are produced by a class of lymphocyte known as B-cells or *plasma cells* (see also Box 11.1, p. 148). To produce a monoclonal antibody (MAB) from an immunized animal, ideally one would isolate the lymphocytes, culture them *in vitro* and clone out the ones producing the antibody of interest. There are two problems with this approach. First, there has been little success in culturing normal B-lymphocytes by the techniques decribed in Chapter 8. Second, to provide an unlimited supply of antibody, an immortal antibody-producing line is required. Surprising as it may seem, immortal monoclonal antibody-producing cells do exist in nature. They occur in the disease state known as *multiple myeloma*, which is a cancer of the B-lymphocytes. In an individual thus afflicted a neoplastic transformation has occurred in a single plasma cell so that the resulting tumour secretes a homogeneous immunoglobulin. This myeloma protein can comprise 95% of the serum immunoglobulin and the cell which produces it can be cloned easily and grown *in vitro*. The myeloma protein for any one afflicted individual is different from the myeloma proteins of all other individuals and thus different clones producing different monoclonal antibodies can be obtained. However, the choice of monoclonal antibody is restricted to those produced by myelomas studied to date. What is needed is a means of extending the range of antibodies produced by myeloma cells and this was achieved in 1975 in a classic piece of work which later earned a Nobel Prize for George Kohler and Cesar Milstein.

Cells in culture can fuse spontaneously and the frequency with which this occurs can be increased by the addition of fusion-promoting agents such as polyethylene glycol. The immediate fusion products are heterokaryons, in which the cell retains both nuclei intact. In many of the heterokaryons the nuclei fuse to generate stable hybrids which express both parental genomes. What Kohler and Milstein did was to fuse normal antibody-producing mouse lymphocytes with a mouse myeloma cell line to generate a *hybridoma*. The hybridoma is derived from a transformed cell and so it has the potential to be grown indefinitely in culture.

At this stage it is important to realize that not just any myeloma cell can be used, but that only ones which have two particular properties are suitable. First, the myeloma cell line used must itself not be capable of secreting antibody; otherwise the hybridoma will secrete a mixture of antibodies. Such cell lines are readily obtained because, although in most cases of multiple myeloma there is an overproduction of a particular immunoglobulin, there are instances where a particular lymphocyte multiplies uncontrollably without antibody production. Second, the myeloma cells used for fusion are specially selected by growing them in the presence of 8-azaguanine. Most of the cells are killed by this technique but a few survive and these resistant cells have a defect in the enzyme hypoxanthine phosphoribosyl transferase (HPRT).

When HPRT-negative cells are grown in a mixture of hypoxanthine, aminopterin and thymidine (HAT medium), the cells will die because they can no longer synthesize DNA (Fig. 10.2). The aminopterin blocks the main pathway for purine and pyrimidine biosynthesis. Normal cells can use the thymidine to make

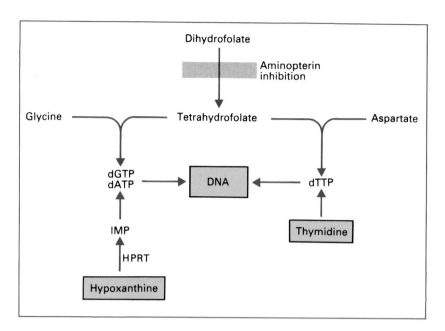

Fig. 10.2 The selective basis for the use of HAT medium. See text for details.

pyrimidines and, using HPRT, can convert the hypoxanthine to purines. In a fusion mixture of lymphocytes and myeloma cells, neither cell can grow in the HAT medium. The lymphocytes will survive for a few days before dying out, as will the myeloma cell, because they lack HPRT. However, the hybridoma cells will possess the ability of the myeloma cells to grow *in vitro* and the normal HPRT gene from the lymphocytes with which they have fused. Thus the hybridoma will grow successfully even in the HAT selection medium.

MAKING A HYBRIDOMA [1]

The starting-point for making a hybridoma which secretes the desired MAB is to immunize an animal, e.g. a mouse, with the appropriate antigen. Normally the antigen is injected subcutaneously or into the peritoneal cavity along with an adjuvant to stimulate the immune system. The animal is injected on several occasions and with each successive immunization there is increased stimulation of the B-lymphocytes which are responding to the antigen. The final dose of antigen is given intravenously 3 days before the animal is killed. The intravenous injection ensures a high dose of antigen and 3 days after immunization the immune-stimulated cells will be growing maximally. This simplifies the selection process later.

After the immunized animal has been killed its spleen is aseptically removed and gently disrupted to release the lymphocytes and red blood cells. The lymphocytes are separated from the red blood cells and splenic fluid by density gradient centrifugation. After washing, the lymphocytes are mixed with an HPRT-negative myeloma cell line. The mixture of cells is exposed to the fusion-promoting agent polyethylene glycol, but only for a few minutes since it is cytotoxic. The cells are then washed free of the polyethylene glycol by suspending the pellets in fresh medium. The washed cells comprise a mixture of hybridomas, unfused myeloma cells and unfused lymphocytes. By using HAT growth medium the hybridomas can be selected as described above.

If all the hybridomas that occurred after a fusion were grown together, then a polyclonal antibody mixture would be obtained. Consequently, single antibody-producing hybridoma cells need to be isolated and grown individually. This is done by diluting a suspension of hybridoma cells to such an extent that individual aliquots contain, on average, only one cell. The cells are then grown in fresh medium. One problem is that animal cells do not readily grow in isolation but ways of overcoming this are described in Chapter 8 (see p. 115). Each clone is then examined to determine if it produces the desired MAB. Here one encounters a problem of logistics. Even with the best immunization schedule less than 5% of the splenocytes will be producing antibodies and, of those that are, only a small proportion will have responded to the antigen used for immunization. Consequently a mass screening procedure needs to be undertaken and success is not guaranteed! The various steps involved in making a hybridoma are summarized in Fig. 10.3.

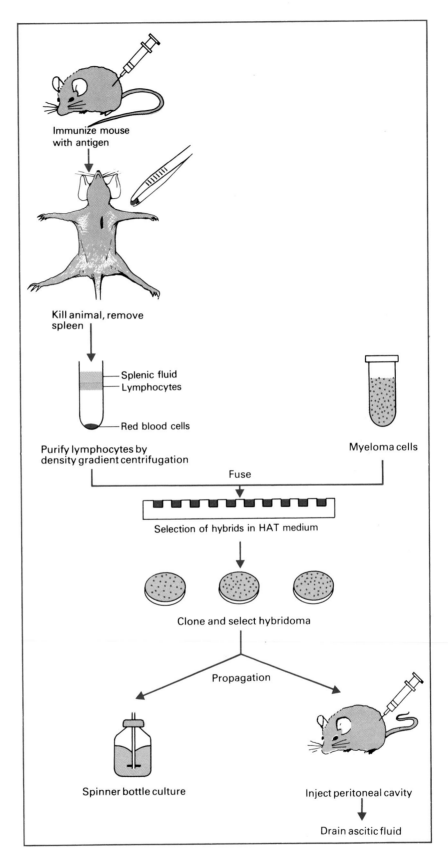

Fig. 10.3 Steps in the production of a hybridoma.

Once the correct hybridoma has been selected, it can be stored frozen and cultured whenever required. Since the hybridoma is a transformed cell line it grows readily in culture but the antibody titre is low. In roller bottle culture antibody levels of 5–10 mg/l of culture medium are not uncommon. One solution is to grow the hybridoma to high cell density in suspension culture with or without the use of microcarriers. In this way antibody levels of 10–100 mg/l can be obtained. An interesting method patented by the Damon Corporation is to grow the hybridomas inside hollow microspheres which are surrounded by a porous membrane (Fig. 10.4). Protected inside the microsphere the cells multiply and grow to higher cell densities than is possible with any other culture system. Since the antibody is retained by the membrane it does not become contaminated with other immunoglobulins if serum has to be included in the growth medium.

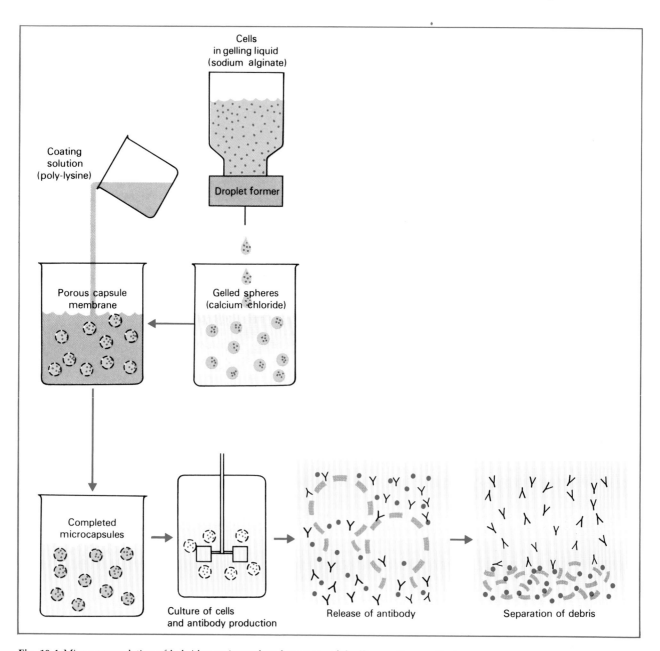

Fig. 10.4 Microencapsulation of hybridomas (reproduced courtesy of the Damon Corporation).

Using this microencapsulation technology 100 mg to 1 g of antibody can be produced per litre of culture medium and this makes large-scale production of MABs commercially feasible.

An alternative way of obtaining a high concentration of antibody is to grow the hybridoma as ascites, i.e. as a suspension of tumour cells within the peritoneal cavity of an isogenic species. The ascitic fluid which forms following injection of the tumour can contain as much as 20 g/l of MAB, a level superior to that obtained by the best *in vitro* cultivation methods. Some commercial suppliers of MABs use the ascites route but it has its disadvantages. Not only is it labour-intensive, but specific pathogen-free (SPF) animals need to be used and kept quarantined at all times to prevent contamination of the MAB. This adds greatly to the cost. In addition, why sacrifice large numbers of laboratory animals when an alternative is available?

Whatever the method used to grow the hybridoma, the MAB produced is usually purified prior to use. If the hybridoma has been grown *in vitro*, purification of the MAB will be facilitated if the culture medium contained only serum albumin and not complete serum. After precipitation with ammonium sulphate the MAB can be purified by ion exchange chromatography, e.g. on a DEAE-based matrix (Fig. 10.5).

HAZARDS ASSOCIATED WITH MABS

The majority of mammalian cells contain in their chromosomes genetic information related to retroviruses. Mice, and in particular the inbred strains used in laboratory studies, carry a number of different types of these viruses and murine cell lines frequently release virus. Retrovirus production has been noted in a number of myeloma cell lines and in hybridomas prepared from them. It is possible to find myeloma lines that do not actively produce virus and, although their use does not guarantee that a hybridoma will be virus-free, it does reduce the risk of contamination. Due to the dangers which those viruses can pose, regulatory authorities have stipulated that medicinal products containing MABs or produced using MABs

Fig. 10.5 Purification of a MAB from ascites fluid. The main figure shows the separation of IgG from contaminating proteins by ion exchange chromatography of ascitic fluid. The inset shows an electropherogram of crude ascitic fluid (track 1) and various IgG fractions (tracks 2–4).

will not be approved unless the MABs and/or the hybridomas can be shown to be virus-free. Table 10.1 lists those viruses of concern to the US Food and Drug Administration: until the advent of the PCR (p. 24) demonstrating their absence was not easy! An alternative method of producing MABs, and one which eliminates the virological hazards, is to synthesize them in bacteria using recombinant DNA technology. This is discussed later.

Applications of MABs [2]

For many years polyclonal antisera produced by bleeding immunized animals have been widely used for diagnostic purposes, particularly in clinical microbiology and experimental biology. Monoclonal antibodies are far more versatile because of their specificity and high purity and hence they are used in such diverse applications as human therapy and commercial protein purification. As a consequence a whole new industry has grown up and in recent years several biotechnology companies have been founded whose income comes mostly from the sale of MABs.

As reference to Table 10.2 will show, a detailed description of all the uses of MABs is outside the scope of this book. A few representative examples are given below and those wishing further information should consult the cited references.

IMPROVED DIAGNOSTIC REAGENTS

Antibody—antigen interactions are used widely for the assay of a wide range of compounds including hormones, antibiotics and biological effector mole-

Table 10.1 Viruses which are potential contaminants of MABs produced by hybridomas of murine origin.

Adenovirus
Encephalomyelitis virus
Encephalomyocarditis virus
Lymphocytic choriomeningitis virus
Hepatitis virus
Polyoma
Retroviruses
Sendai virus
Thymic virus
Reovirus
Haemorrhagic fever virus
Ectromelia virus
Lactate dehydrogenase virus
Cytomegalovirus

Table 10.2 Some of the applications of MABs.

Improved sensitivity and reproducibility of existing immunoassays or new assays for:
Histocompatibility antigens
Fibronectin
Blood group antigens
Sperm antigens
Interleukins
Complement components
Interferons
Progesterone
Gastrin
Blood clotting factors
Oestrogen
Human growth hormone

Diagnosis of:
Sexually transmitted diseases
Cancer (by diagnostic imaging)

Therapy:
Correction of drug overdose
Reduction of risks associated with bone marrow transplants
Detection of tumour metastases
Treatment of cancer (directly or by targeting cytotoxic drugs)

Catalysis:
Conversion of substrates

cules such as interferons and blood clotting factors. Also, the ability of antibodies to agglutinate or precipitate cells has long been used in diagnostic microbiology and in blood typing. Prior to the development of MABs polyclonal antisera were used but these have a number of disadvantages. For example, serum batches from different animals or groups of animals, or even one animal at different times, have different antibody contents and usually do not show an identical reaction pattern. Thus there will be considerable batch to batch variation in the diagnostic reagent and this generates a requirement for repeated calibration. This inherent variability does not occur with MABs since they should be produced in a consistent fashion by repeated subculture of a master cell bank of a hybridoma. Another problem is that polyclonal antisera contain only a small proportion of the antibody of interest. Since the rate of interaction between an antigen and an antibody is concentration-dependent this means that reaction times are prolonged. The homogeneity of MABs reduces reaction times and there is less likelihood of non-specific cross-reactions occurring. As a consequence of these advantages — consistency, specificity and speed — many diagnostic kits based on MABs are now on the market.

Monoclonal antibody affinity columns are readily prepared by coupling MABs to a cyanogen bromide-activated chromatography matrix, e.g. Sepharose. MABs immobilized in this way are particularly valuable for the purification of proteins. There are at least three advantages of immunoaffinity methods over conventional purification techniques. First, since the MAB has a unique specificity for the desired protein the level of contamination by unwanted protein species is usually very low. Second, since the MAB—antigen complex has a single binding affinity it is possible to elute the required protein (antigen) in a single, sharp peak (Fig. 10.6). Third, the concentration of the required protein relative to total protein in a mixture is irrelevant since the capacity of the antibody remains unchanged. The data in Table 10.3 show that 97% recovery of interferon was obtained by immuno-affinity chromatography when the starting concentration was less than 0.02%. The fact that a 5300-fold purification was obtained in a single step highlights the power of the method and it is not surprising that it is being used commercially to purify recombinant-derived interferon-α_2. However, a word of caution is necessary: 100% pure protein is difficult to achieve by this method because there is always a tendency for small amounts of immunoglobulin to leak off the immunoaffinity column. Furthermore, MABs often do not distinguish between intact protein molecules and fragments containing the antigenic site.

BONE MARROW AND ORGAN TRANSPLANTATION [4, 5]

There are several diseases where the injection of bone marrow from a healthy individual may be of considerable benefit to a patient. These include a wide variety

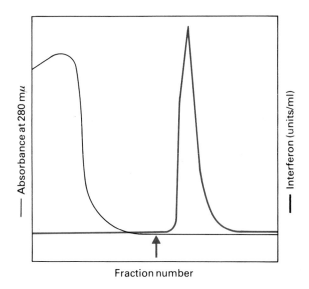

Fig. 10.6 Purification of interferon-α by affinity chromatography. Crude interferon was loaded on to a column of anti-interferon MAB immobilized to Sepharose. After loading the unbound protein was eluted by washing with buffer. At the point shown by the arrow, an acid-wash (pH 2.0) was started to elute the interferon.

of congenital marrow aplasias where erythrocyte stem cells stop dividing, resulting in severe anaemia. These aplasias can be cured if the red cells in the marrow can be replaced by normal stem cells. Disorders of leucocyte and lymphocyte stem cells also occur, resulting in greatly decreased resistance to microbial infection.

There are two major problems associated with bone marrow transplantation. The first of these is the rejection of the donor cells by the host. Eventually the use of MABs will permit better typing of donor and recipient lymphocytes to enable better cross-matching. In the mean time, immunosuppressive therapy is required for a prolonged period of time after transplantation but there is a consequently higher risk

Table 10.3 Purification of crude interferon from tissue culture medium using an immobilized anti-interferon antibody (adapted from Secher & Burke 1980).

	Before	After
	Immunoaffinity chromatography	
Volume of interferon-containing solution	100 ml	0.5 ml
Interferon titre (u/ml)	7.2×10^4	1.5×10^7
Total units	7.2×10^6	7.0×10^6
Specific activity of interferon (u/mg)	3.3×10^4	1.8×10^8
Purification factor	1	5300
Recovery of interferon	100	97
Size of immunoaffinity columns	0.5 ml	

of mortality from Gram-negative bacteraemia. The problem stems from the fact that even if antibiotics successfully combat the infection, the released endotoxin can still cause death. A polyclonal anti-endotoxin antibody has been produced in humans and shown to be effective in reducing the death from bacteraemia. More recently a human MAB against *E. coli* endotoxin has been produced which protects mice against lethal Gram-negative bacteraemia. Not only is this MAB effective in humans, it can be shown to be free from hepatitis, cytomegalovirus and HIV, which are potential contaminants of human polyclonal antisera.

A second problem associated with marrow transplants is that the marrow contains donor T-lymphocytes. These can recognize cells in the new host as being foreign and begin to destroy them. This produces the clinical syndrome known as graft-versus-host (GVH) disease, which can lead to death although it is partly controlled by immunosuppressive drugs. An anti-T-cell monoclonal antibody is available and has been used to remove T-cells from the donor marrow prior to transplantation, leading to a reduction in GVH disease. A second use for the anti-T-cell monoclonal antibody, which has received regulatory approval, is to prevent rejection of donor organs such as heart, liver and kidneys after transplantation. Following administration of the antibody the T-cells are removed from the circulation and the patient temporarily becomes immunologically incompetent. A foreign organ, such as a kidney, introduced into the body at this time is not recognized as being antigenic, so antibodies to it are not synthesized when T-cells reappear in the bloodstream.

One side effect of the use of such antibodies is an increase in lymphomas among transplant recipients. Apparently, when T-cells become overly inhibited, the dormant viruses which most people harbour flare up in an infection. This in turn triggers a proliferation of B-cells which goes unchecked by T-cells. Any cancerous B-cells which arise rapidly result in a lymphoma.

IMMUNOTOXINS AND RADIOIMMUNOTHERAPY [6, 7, 8]

Many tumour cells possess additional or new antigens on their surface in comparison with the normal cells from which they are derived. When tumour cells are used as immunogens some of the resulting antibodies bind only to tumour-specific antigens. Such antibodies can be used for cancer therapy by coupling to them either toxins (*immunotoxins*) or, in the case of *radioimmunotherapy*, radioisotopes. Another immunological approach to the treatment of cancer, but which does not involve antibodies, is *adoptive immunotherapy* (see Box 10.1).

Immunotoxins. A number of different toxins or toxin fragments have been coupled to antibodies, e.g. diphtheria toxin, *Pseudomonas* exotoxin, etc. The example chosen here is ricin. Ricin is a protein derived from the castor oil plant and which is cytotoxic. It consists of two peptide chains linked by a disulphide bond. The B-chain binds to galactose residues on the surface of sensitive cells and this binding facilitates the entry of the A-chain. Once the A-chain gains access to the cell cytoplasm it enzymatically and irreversibly alters the ribosomes so that they can no longer function. This leads to cell death. The toxic A-chain of ricin can be separated from the B-chain and coupled to the MAB. This newly created immunotoxin should bind to tumour cells and not normal cells (Fig. 10.7) and lead to death of the former. Since the A-chain of ricin inhibits protein synthesis and not cell division it kills even those cells which are potentially harmful but not actively dividing.

There are many problems to be solved before immunotoxins are successful. The ability of immunotoxins to target tumour cells amidst a background of normal tissues depends to a large extent on the specificity of the antibody. However, the host immune clearance mechanisms will try to redirect some of the conjugate to immune cells or the liver and this may have to be reduced with immunosuppressive drugs. Tumour cells may also shed antigens into the blood. Not only will these free antigens mop up the immunotoxin, but they too will redirect the immunotoxin to the liver. This problem may be overcome by plasmapheresis of patients before treatment to remove such antigens. Finally, to be effective, the immunotoxins will need to penetrate the tumour mass and this might require the use of smaller conjugates derived from immunoglobulin fragments.

Radioimmunotherapy (RAIT). An alternative approach to the use of immunotoxins is to couple a radioisotope to the MAB such that the tumour cell is killed by irradiation. One advantage of this method is that it reduces the need for the immunoconjugate to penetrate the tumour mass. For the purpose of RAIT the isotope

Adoptive immunotherapy, also called cell-transfer therapy, is one of the new class of approaches being developed to strengthen the innate ability of the immune system to fight cancer. There are two basic methods and both rely on the properties of a molecule called interleukin-2 (or T-cell growth factor). This cytokine is produced naturally by a class of cells known as helper T-cells but commercially it is prepared by recombinant DNA technology (see Chapter 6). Interleukin-2 induces helper T-cells and antigen-stimulated cytotoxic T-cells to replicate.

In LAK therapy, lymphocytes are isolated from the blood of patients and cultured for 3 days with interleukin-2. This causes certain lymphocytes to become LAK cells (lymphokine-activated killer cells). About 50 billion of these LAK cells are infused intravenously into the patient along with interleukin-2. Although the side-effects of

the therapy are horrendous, complete regression has been obtained in a high proportion (20%) of patients with certain advanced cancers, e.g. cancer of the kidney, melanoma and non-Hodgkin's lymphoma.

In TIL therapy, a nodule of cancerous tissue is removed from the patient and digested with enzymes to produce separated cells. These cells are incubated with interleukin-2. During this period certain lymphocytes from the tumour, the tumour-infiltrating lymphocytes (TILs), begin to proliferate rapidly and to attack the cancer cells. After 30—45 days the lymphocytes in the culture completely replace the tumour cells and are infused into the patient along with more interleukin-2. TILs are more potent than LAK cells and the tumour regression rate is higher, but again the side-effects of the interleukin-2 are particularly unpleasant for the patient.

---

selected needs to have high energy (to kill target cells), low penetration (to prevent damage to surrounding tissue) and a short half-life (so that the patient is not a radiation hazard!) and the products of radioactive decay should be inert. The radioisotope which comes closest to fulfilling these needs is yttrium-90, which is a β particle emitter with a half-life of 64 hours. Unfortunately it is only just becoming available in sufficient quantities and so most experiments to date have been done with less suitable isotopes, e.g. indium-111. The early clinical results with this technique have been very encouraging.

DIAGNOSTIC IMAGING

As well as being used for cancer therapy, radiolabelled antibodies can be used to diagnose certain disease states, including cancer. For example, MABs to fibrin can be used to locate blood clots associated with deep vein thrombosis and tumour-specific MABs to locate primary or metastasized cancers. The appropriate MAB is administered intravenously and accumulates at sites of antigen concentration. These sites are de-

tected with the aid of such devices as gamma cameras. Clearly, in terms of emission characteristics, the requirements for the radiolabel used are different from those for radioimmunotherapy. The isotope of choice is technetium-99m.

CATALYTIC ANTIBODIES [12, 13, 14, 15]

In many ways, enzymes are ideal catalysts. They display extraordinary substrate specificity and reaction selectivity and operate at near neutral pH values and at low temperatures. Unfortunately, for many applications enzymes with the required substrate or product specificity are not known. Antibodies are virtually unlimited in their range of recognition site structure and may provide the required catalytic agents.

Antibodies, like enzymes, are proteins which bind other molecules (antigens) with high affinity and specificity. The interactions of the antibody binding domain with antigen are similar to those of an enzyme with substrates, e.g. electrostatic, hydrogen and van der Waals bonding. If a molecule, resembling in conformation and charge the transition state between sub-

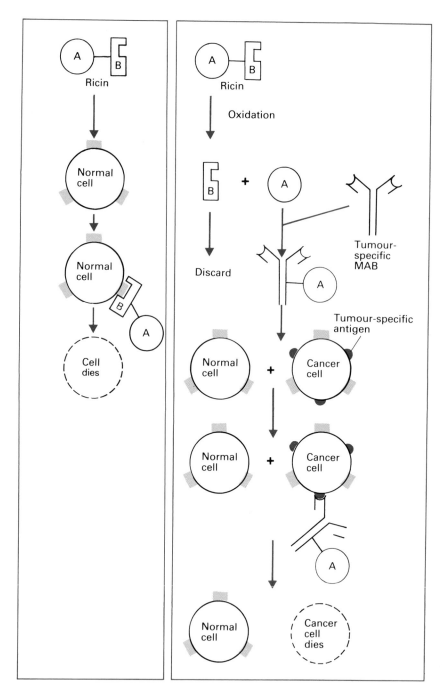

Fig. 10.7 Use of immunotoxins as therapeutic agents.

strate and product, is stable enough to be used as a hapten, the resulting antibody should be capable of catalysis. That this is indeed the case has been shown for a whole variety of reactions (Table 10.4). High specificity, including enantiomer selectivity, and rate enhancements up to 10^7-fold have been observed.

It is doubtful whether catalytic MABs will ever match

Table 10.4 Examples of reactions catalysed by antibodies.

Hydrolysis of esters, amides and carbonates
Biomolecular reactions
Cyclization reactions
Elimination reactions
Asymmetric synthesis

the catalytic efficiency of an enzyme for its natural substrate, although site-directed mutagenesis has enabled improved variants to be produced. Thus it is likely that their eventual applications will be those where natural enzymes do not have the desired specificity or where a known enzyme does not exist.

THE GENETICS OF ANTIBODY DIVERSITY [16]

In recent years a number of *second-generation* MABs with improved properties have been produced. Also, more rapid methods of isolating MABs have been developed. To understand how these advances have been made it is necessary to know how antibody diversity is generated.

Germ line cell DNA contains many regions encoding parts of antibody molecules. These coding regions are brought together by a combination of DNA recombination and RNA splicing. During the development of antibody-producing cells (B-cells) one V (variable) and one J (join) gene segment recombine to produce a VJ gene. For heavy chains, a D (diversity) segment is also incorporated to form a VDJ gene. A primary RNA transcript produced from the functional gene includes sequences encoding the constant region. Introns separating the various sequences are spliced out as the RNA is processed to messenger. Heavy and light chains are produced in the same cell from genes on different chromosomes and, after synthesis, combine to produce a multichain molecule (Fig. 10.8).

The diversity of immunoglobulin specificity is due to the presence of hundreds of V genes, recombination between V, D and J genes during B-cell maturation, inaccuracies in recombination, point mutations in immunoglobulin genes encoding hypervariable regions and the combination of different light and heavy chains. There is also diversity in the antibody constant region. The light chains are either kappa or lambda type. There are four lambda constant regions but only one kappa C (constant) gene. The various heavy chain constant region genes determine the immunoglobulin class and subclass (e.g. IgM, IgG, etc.).

HUMANIZED MABS [16, 17]

Although rodent monoclonal antibodies have been used in a number of clinical situations, the patient's immune system recognizes the administered antibodies as foreign and synthesizes anti-antibodies, e.g. the so-called HAMA (human anti-mouse antibody)

response. This is of particular concern when clinical protocols necessitate multiple applications of antibody reagents. The use of human MABs would be much more satisfactory but there are ethical restraints on the necessary human immunization! To partially address this problem several laboratories have produced chimeric antibodies, in which the antigen-binding variable region from a rodent source is genetically combined to a human constant region (Fig. 10.9). Such antibodies have been further 'humanized' by eliminating the rodent framework regions (see p. 128) from the variable domain of the heavy chain.

The humanizing of MABs has only been possible because of advances in recombinant DNA technology. The polymerase chain reaction has been particularly helpful. Despite the great diversity of antibody genes, all light and heavy chain mRNA molecules have conserved regions. Primers to those conserved regions allow PCR amplification and gene isolation. The cloned genes can then be manipulated *in vitro* and transformed into bacteria, yeasts or animal cells, which secrete the modified MABs. The antibodies can be further modified by site-directed mutagenesis to enable cysteine residues to be introduced at predetermined locations on the antibody surface. Such cysteine residues facilitate the covalent labelling of antibodies with effector or reporter molecules for use in radioimmunotherapy and diagnostic imaging.

SINGLE DOMAIN ANTIBODIES [18, 19]

Both the heavy and light chain variable domains contribute to the binding of antigen molecules. There is some asymmetry, for the light chain contributes less than 50% of the antigen-binding surface. Nevertheless, it was surprising when a group of workers were able to show that isolated V_H domains prepared from bacteria bind antigens with an affinity only 10-fold less than that of the parent antibody. These antigen-binding domains have been given the name *single domain antibodies* (dABs).

To prepare dABs, mice are immunized with an antigen and their spleens removed. Using the polymerase chain reaction for amplification, diverse libraries of V_H genes are cloned from the spleen genomic DNA. The cloned genes are introduced into *E. coli*, which secrete the desired heavy chain variable domains. An advantage of this technique is that tissue culture and hybridoma selection are avoided. Rather, antigen-binding activity from *E. coli* can be obtained

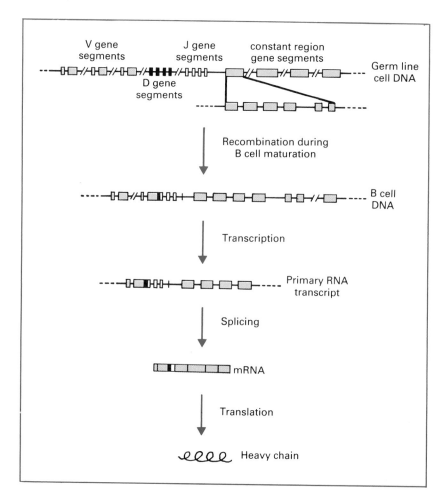

Fig. 10.8 The genetics of antibody synthesis. The figure shows the sequence of events which results in the synthesis of an antibody heavy chain. The sequence of events (not shown) occurring in the synthesis of light chains is simpler since there are no D gene segments and fewer constant region gene segments. After synthesis, the heavy and light chains combine before secretion.

within 3 days of harvesting the spleen of the immunized animals. An even simpler procedure is to express the antigen-binding activity on the surface of a bacteriophage. Those phages that express the V_H genes can be isolated by affinity chromatography using immobilized antigen.

Compared with intact antibodies, dABs have reduced affinities for antigens and are unlikely to replace MABs in the many applications discussed earlier. However, they could be useful in applications such as immunotoxin therapy where their small size should facilitate tissue penetration.

Single chain antibodies (SCAs) [21, 22, 23,] are artificial antibodies composed of heavy- and light-chain binding regions linked chemically and produced in bacteria. To make an SCA, the gene sequences for the N-terminal end of the heavy and light chains are joined by a gene sequence encoding a linker peptide and the entire construct inserted into an appropriate expression vector. The advantage of SCAs over dABs is that the former combine small size, to facilitate tissue penetration, with good affinity for antigens.

FAB LIBRARIES [20]

An extension of the dAB technology is to clone not just V_H genes but the genes for the entire light chain and the corresponding part (Fd) of the heavy chain. The technology used is similar but the product is Fab molecules. Like dABs, Fabs can be made in a matter of days without cell culture but they have the added benefit of better antigen binding. Furthermore, the actual combinations of V_C and V_H domains are random and do not necessarily represent the combinations present in the B-cell population in the parent animal. This could be particularly beneficial when looking for catalytic antibodies.

Fig. 10.9 Schematic representation of humanized antibodies. The solid bars represent sequences derived from rodent genes. In (a) the entire variable region is of rodent origin. In (b) only the complementarity-determining region is of rodent origin.

Further reading

GENERAL REFERENCES

Hilvert D. (1991) Extending the chemistry of enzymes and abzymes. *Trends in Biotechnology* **9**, 11–17.

Sikora K.& Smedley H.M. (1984) *Monoclonal Antibodies*. Blackwell Scientific Publications, Oxford.

Gorman C.M. (1990) Mammalian cell expression. *Current Opinion in Biotechnology* **1**, 36–47.

Winter G. & Milstein C. (1991) Man-made antibodies. *Nature* **349**, 293–99.

Powell M.J. & Hansen D.E. (1989) Catalytic antibodies — a new direction in enzyme design. *Protein Engineering* **3**, 69–75.

CITED REFERENCES

1 Kohler G. & Milstein C. (1975) Continuous cultures of fused cells producing antibodies of predefined specificity. *Nature* **256**, 495–7.

2 Seiles F.S., Gronski P., Kurrle R., Lueben G., Harthus H.P., Ax W., Bosslet K. & Schwick G. (1985) Monoclonal antibodies: their chemistry, functions and possible uses. *Angewandte Chemie (International Edition)* **24**, 139–60.

3 Secher D.S. & Burke D.C. (1980) A monoclonal antibody for large scale purification of human leukocyte interferon. *Nature* **285**, 446–50.

4 Benjamin R.J. & Waldmann H. (1980) Induction of tolerance by monoclonal antibody therapy. *Nature* **320**, 449–51.

5 Teng N.N.H., Kaplan H.S., Herbert J.M., Moore C., Douglas H., Wunderlich A. & Braude A.I. (1985) Protection against Gram-negative bacteraemia and endotoxaemia with human monoclonal IgM antibodies. *Proceedings of the National Academy of Sciences (USA)* **82**, 1790–4.

6 Ahmad A. & Law K. (1988) Strategies for designing antibody–toxin conjugates. *Trends in Biotechnology* **6**, 246–51.

7 Kozak R.W., Waldmann T.A., Atcher R.W. & Gansow O.A. (1986) Radionuclide-conjugated monoclonal antibodies: a synthesis of immunology, inorganic chemistry and nuclear science. *Trends in Biotechnology* **4**, 259–64.

8 Osband M.E. & Ross S. (1990) Problems in the investigational study and clinical use of cancer immunotherapy. *Immunology Today* **11**, 193–5.

9 Russell S.J. (1990) Lymphokine gene therapy for cancer. *Immunology Today* **11**, 196–200.

10 Rosenberg S.A. (1990) Adoptive immunotherapy for cancer. *Scientific American* **262**, 34–41.

11 Rees R.C. & Wiltrout R.H. (1990) The biology and clinical applications of interleukin 2. *Immunology Today* **11**, 36–8.

12 Schultz P.G. (1988) The interplay between chemistry and biology in the design of enzymatic catalysts. *Science* **240**, 426–33.

13 Green B.S. & Tawfile D.S. (1989) Catalytic monoclonal antibodies: tailor-made, enzyme-like catalysts for chemical reactions. *Trends in Biotechnology* **7**, 304–9.

14 Shokat K.M., Leumann C.J., Sugasawara R. & Schultz P.G. (1989) A new strategy for the development of catalytic antibodies. *Nature* **338**, 269–71.

15 Baldwin E. & Schultz P.G. (1989) Generation of a catalytic antibody by site-directed mutagenesis. *Science* **245**, 1104–7.

16 Williams G. (1988) Novel antibody reagents: production and potential. *Trends in Biotechnology* **6**, 36–40.

17 Lyons A., King D.J., Owens R.J., Yarranton G.T., Millican A., Whittle N.R. & Adair J.R. (1990) Site specific attachment to recombinant antibodies via introduced surface cysteine residues. *Protein Engineering* **3**, 703–8.

18 Ward E.S., Gussow D., Griffiths A.D., Jones P.T. & Winter G. (1989) Binding activities of a repertoire of single immunoglobulin variable domains secreted from *Escherichia coli*. *Nature* **341**, 544–6.

19 McCafferty J., Griffiths A.D., Winter G. & Chiswell D.J. (1990) Phage antibodies: filamentous phage displaying antibody variable domains. *Nature* **348**, 552–4.

20 Huse W.D., Sastry L., Iverson S.A., Kang A.S., Alting-Mees M., Burton D.R., Benkovic S.J. & Lerner R.A. (1989) Generation of a large combinatorial library of the immunoglobulin repertoire in phage lambda. *Science* **246**, 1275–81.

21 Bird R.E., Hardmann K.D., Jacobson J.W., Johnson S., Kaufman B.M., Lee S.-M., Lee T., Pope S.H., Riordan G.S. & Whitlow M. (1988) Sing-chain antigen-binding proteins. *Science* **242**, 423–6.

22 Huston J.S., Levinson D., Mudgett–Hunter M., Tai M.-S., Novotny J., Margolies M.N., Ridge R.J., Bruccoleri R.E., Haber E., Crea R. & Opperman H. (1988) Protein engineering of antibody binding sites: recovery of specific activity in an anti-digoxin single-chain Fv analogue produced in *Escherichia coli*. *Proceedings of the National Academy of Sciences US* **85**, 5879–83.

23 Davis G.T., Bedzyke W.D., Voss E.W. & Jacobs T.W. (1991) Single chain antibody (SCA) encoding genes: one step construction and expression in eukaryotic cells. *Bio/technology* **9**, 165–9.

11 Transgenic Animals and Human Gene Therapy

Gene therapy is the correction of an inborn error of metabolism by the insertion into the afflicted organism of a normal gene. Ideally, the inserted gene will be correctly targeted and regulated. This has been achieved in a number of experimental animal systems and it will not be long before it will be possible to cure some human diseases in a similar way. This immediately raises a whole series of ethical questions some of which are discussed later (see p. 188), but suffice it to say that the first attempts are underway.

In the case of animals other than man there is no overwhelming desire to keep alive those individuals with severe genetic impairment. However, the same techniques which are being used for gene therapy could be developed into sophisticated animal breeding programmes either to:

1 introduce new characteristics not available before; or to

2 introduce existing characteristics more quickly than by conventional breeding.

As far as contemporary thought is concerned, the insertion of genetic material into a human being for the sole purpose of correcting a genetic defect in that patient, i.e. somatic cell gene therapy, is socially acceptable. Attempts to modify germ cells or to enhance or 'improve' a 'normal' person by gene manipulation are abhorrent, particularly so in view of the difficulties which would be encountered using current techniques. By contrast, in animal breeding, modification of germ cells and 'improvement' of the animal are the desired goals. As will be seen later, the modification of germ cells by gene manipulation *in vitro* is possible and this process is given the name *transgenesis*. Transgenesis is not restricted to animals but can also be achieved in plants (see Chapter 13).

GENE DELIVERY METHODS

Regardless of whether somatic or germ cells are to be modified, a method of introducing exogenous DNA into those cells is required. Table 11.1 lists the methods currently available. The first two methods have been dealt with in detail in Chapter 9 and will not be described again here. Microinjection is a widely used technique but is a laborious procedure as only one cell at a time can be injected; it also has a high failure rate (see, for example, Table 11.3, p. 145) because even in the hands of experienced operators many of the eggs are damaged. Electroporation and cell fusion are not used much because little is known about the mechanics of the processes but they would find favour if they could be shown not to result in damage to the recipient cells.

None of the delivery methods listed in Table 11.1 fulfils the criterion of being able to transfect the majority of cells with the minimum cellular damage/mortality. Where somatic cells are to be modified the use of viral vectors or calcium phosphate-mediated uptake is indicated; for modification of germ cells microinjection is the method of choice. Whatever the DNA delivery method used one problem still remains: there is no control over where the donor DNA will integrate in the recipient genome and whether this will result in any pathological condition.

PRODUCTION OF TRANSGENIC MICE [1, 2]

As indicated above, the aim of transgenesis is to produce animals with a hereditable change in their genotype such that the benefits of gene manipulation can be passed to their offspring. Economically the current interest is in the modification of farm animals but in the future attention may be paid to other food

Table 11.1 Methods of introducing exogenous DNA into mammalian cells.

1 Use of viral vectors
2 Calcium phosphate-mediated DNA uptake
3 Microinjection of eggs
4 Fusion of DNA-loaded membranous vesicles (e.g. liposomes) to target cells
5 Electroporation (DNA uptake mediated by electric current)

sources such as fish. With mammals, the technique used is the microinjection of DNA into fertilized eggs followed by implantation of the manipulated ova in a foster mother. For *experimental* purposes the mouse is a particularly appropriate choice of animal: a super-ovulated mouse can yield up to 40 eggs, reimplantation is relatively easy and a mouse can carry up to 20 offspring.

In practice, about 2 pl of buffer containing cloned plasmid DNA is injected into one of the pronuclei of the newly fertilized egg (Fig. 11.1). The male pronucleus, derived from the sperm, is larger than the female pronucleus and so is chosen for injection. The

Fig. 11.1 Injection of DNA into the pronucleus of a newly fertilized egg.

two pronuclei subsequently fuse to form the diploid zygote nucleus of the fertilized egg. The injected embryos are cultured *in vitro* until they have undergone a number of divisions and are then implanted in pseudopregnant foster mothers. Between 3% and 40% of the animals developing from these embryos contain copies of the exogenous DNA integrated into their chromosome. In any particular experiment the copy number of the integrated plasmid sequence in individual transgenic animals can vary over a hundredfold range and the chromosomal location differs.

Most of the early experiments on transgenesis made use of the mouse metallothionein (MMT) gene promoter. MMT encodes a small cysteine-rich polypeptide which is involved in heavy metal homeostasis. As such it is inducible not only by heavy metals but also by glucocorticoid hormones. Although distributed in many tissues it is most abundant in the liver. By coupling the MMT promoter to a herpes simplex thymidine kinase (TK) gene it is possible to raise transgenic mice in which TK activity shows a tissue distribution similar to MMT and is inducible by cadmium ions. This experiment shows that it is possible for a foreign gene, in this case from a virus, to be integrated into the genome of a mouse. More important, the TK gene is inherited by offspring as though it were integrated into a single chromosome. However, the level of expression in the offspring varies considerably from that of the parent, some progeny having increased expression and others decreased expression.

A more spectacular example is the coupling of the MMT promoter to a rat growth hormone gene. Transgenic mice carrying this gene construct grow significantly faster and larger than their litter-mates which do not (Figs 11.2 and 11.3). In some cases the transgenic mice grow to almost double the weight of normal mice even when the recipient mice are homozygous for growth hormone deficiency. Again, the donor gene is inherited in a Mendelian fashion.

In an ideal transgenesis experiment the promoter selected would be one which enables the normal cellular control patterns to be exerted on the gene of interest, i.e. the gene should be switched on and off at the correct time and in the correct tissues. Success of this kind has been achieved with transgenic mice carrying the human insulin gene. Not only was gene expression restricted to the islet cells of the mouse pancreas but serum levels of human insulin were properly regulated by glucose, amino acids and an oral hypoglycaemic agent (tolbutamide).

Fig. 11.2 Two male mice from transgenesis experiments. The two mice are siblings and approximately 10 weeks old. The mouse on the left contains a new gene composed of the MMT promoter fused to the rat growth hormone structural gene. The male with the new gene weights 44 g and his sibling without the gene weighs 29 g. The gene is passed on to offspring, which also grow larger than controls. In general, mice that express the gene grow two to three times as fast as controls and reach a size up to twice normal. (Photograph courtesy of Dr R. Brinster, University of Pennsylvania.)

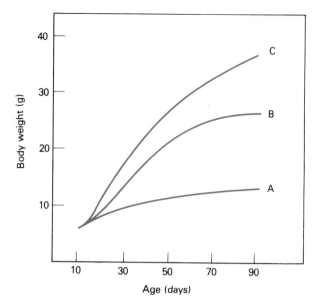

Fig. 11.3 Effect of rat growth hormone on growth of mice. Curve A shows the growth of mice homozygous for a defect resulting in the absence of murine growth hormone. The growth of transgenic litter-mates carrying multiple copies of a rat growth hormone is shown in curve C. Note that the rate of growth and final size of the transgenic mice is greater than that of wild-type mice (curve B). (Adapted from Hammer *et al.* 1984.)

OVERPRODUCTION OF PROTEINS IN TRANSGENIC MICE [3, 4]

Based on the work with rat growth hormone, transgenic mice have been generated which overexpress foreign proteins and secrete them in their milk. This was achieved by coupling the gene of interest to promoters derived from genes encoding milk proteins, e.g. casein. Proteins synthesized in this way include sheep β-lactoglobulin, human tissue plasminogen activator and human urokinase. None of the sheep exhibited any abnormalities even when their milk contained urokinase at concentrations as high as 2 mg/ml. Such experiments have been done to provide models for the eventual production of transgenic farm animals producing high levels of recombinant proteins in their milk. If such proteins were intended for human therapeutic use it is doubtful if regulatory authorities would consider transgenic animals as a reliable production source!

It will be recalled from an earlier chapter (p. 120) that tumour lines can be found which overproduce certain proteins of interest but that the range of lines available is very restricted. Transgenesis provides a means of extending this range. Gene constructs are made in which the gene of interest and an oncogene (tumour gene) are put under the control of a promoter which only functions in lymphoid cells. Transgenic animals carrying such constructs develop lymphoid tumours which overproduce the foreign protein. Cells from these tumours can be used to establish permanent cell lines which can be cultured *in vitro*.

TRANSGENIC MICE WITH HUMAN DISEASE EQUIVALENTS [5, 6]

Currently there are many human diseases for which animal models are lacking and this greatly hinders the development of therapeutic agents. By means of transgenesis a number of mice have been produced which carry genetic lesions identical to those existing in certain human inherited diseases (Table 11.2). Not only will such mice allow a full understanding of the biochemical consequences of gene deficiency but they should also make possible the preliminary screening of potential drugs.

Table 11.2 Examples of mice with human disease equivalents produced by transgenesis.

Genetic alteration	Method of genetic alteration	Human disease equivalent
Introduction of mutant collagen gene into wildtype mice	Nuclear microinjection of inducible minigene	Osteogenesis imperfecta
Inactivation of mouse gene encoding hypoxanthine-guanine phosphoribosyl transferase (HPRT)	Insertion of retrovirus into HPRT locus in embryonic stem cells	HPRT deficiency
Mutation at locus for X-linked muscular dystrophy	Male mutagenesis followed by identification of female carriers	X-linked muscular dystrophy
Introduction of activated human *ras* and *c-myc* oncogenes	Nuclear microinjection of inducible minigene	Induction of malignancy
Introduction of mutant (Z)allele of human α_1-antitrypsin gene	Microinjection of DNA fragment bearing mutant allele	Neonatal hepatitis
Introduction of HIV *tat* gene	Microinjection of DNA fragment	Kaposi's sarcoma
Introduction of β-globin sickle gene	Microinjection	Sickle-cell anaemia
Introduction of mouse renin gene[a]	Microinjection[a]	Hypertension

[a] Transgenic rats produced from fertilized rat eggs.

TRANSGENESIS IN LARGE MAMMALS [7, 8, 9, 10, 11]

While transgenic mice are of interest as an experimental tool, the technique needs to be applied to farm animals if any commercial benefit is to ensue. Working with large domestic animals is much more difficult than with mice. First, they do not produce as many eggs, e.g. a ewe on average only gives three or four. Second, reimplantation of manipulated embryos is more difficult because sheep and cattle will not give birth to more than two offspring. Finally, the eggs of many domestic animals have such an opaque cytoplasm that it is impossible to see their pronuclei or nuclei without resorting to special techniques. One technique which has been used to visualize the pronucleus in pig embryos is centrifugation to separate the cytoplasmic contents. This centrifugation step does not appear to affect their subsequent development. The difficulties cited above are evident from the data in Table 11.3 which show the results obtained when attempts were made to isolate transgenic rabbits, sheep and pigs.

Table 11.3 Efficiency of production of transgenic animals by microinjection of a growth hormone gene. (Adapted from Hammer *et al.* 1985.)

Animal species	No. of ova injected	No. of offspring	No. of transgenic offspring
Rabbit	1907	218	28
Sheep	1032	73	1
Pig	2035	192	20

The results obtained when large animals overexpress growth hormone have been far from satisfactory. Transgenic sheep overproducing growth hormone do grow leaner and have an increased feed efficiency but are more susceptible to infection, tend to die young and the females are infertile. In their pig counterparts the same beneficial effects were seen: improved daily weight gain and feed efficiency and reduced subcutaneous fat. However, the adverse effects were even more dramatic than in sheep; a high incidence was observed of gastric ulcers, arthritis, cardiomegaly, dermatitis and renal disease. Perhaps what is needed is better control of gene expression, a different genetic background or a modified husbandry regimen. The results with human insulin expression in mice suggest that ultimately success will be achieved with animals of commercial value. Certainly, the choice of gene is important because transgenic sheep which secrete human factor IX or α_1-antitrypsin in their milk appear to be physiologically normal and are fertile. Similarly, a fertile transgenic cow has been developed which expresses human interferon-β. Since interferons are very species-specific no adverse effects were expected.

The goal of transgenesis is *not* to produce cows as big as elephants but to introduce specific, economically significant traits into livestock. What is envisaged is that tomorrow's transgenic animals will be more efficient at utilizing feed, will have leaner meat, will grow to marketable size sooner, and will be immune to those diseases that today decimate populations and profits. An encouraging start has already been made in attempts to increase wool production. The keratin proteins of wool are highly cross-linked by disulphide bridges and therefore the demand for cysteine for keratin synthesis in sheep is high. Infusion by cysteine alone, or its precursor methionine, will increase wool growth by up to 100%. Clearly, cysteine is normally available only at suboptimal levels. This is because sulphur is lost as sulphide during the fermentative breakdown and resynthesis of cysteine into proteins by the microflora in the rumen. The microflora are digested later in the abomasum and intestine. One way to overcome this loss would be to produce transgenic sheep containing becterial genes for cysteine synthesis. These genes would be modified such that their expression in the epithelial walls of the upper gastrointestinal tract would enable the liberated sulphide to be 'trapped'. Only two enzymes are required for the reaction (Fig. 11.4) and attaining activity in the ruminal wall would appear to be a sensible target

Fig. 11.4 The biosynthesis of cysteine in sheep rumen. The hydrogen sulphide which is required is produced by fermentative breakdown of food.

since sulphide content is highest in the rumen. Microinjection of the relevant bacterial genes into sheep ova resulted in one transgenic animal. In this animal, expression of the genes could be detected in tail tissue but it is too soon to know the effect on wool production.

GENE TARGETING [12]

Because of the difficulties associated with microinjection of mammalian eggs it is likely that this technique will be superseded by gene targeting (Fig. 11.5). Gene targeting permits the modification of any gene in a defined manner and has the potential to be applied to any animal from which embryonic stem (ES) cells can be obtained.

Gene targeting begins when a targeting vector, containing the desired mutation, is introduced into ES cells (Fig. 11.5). In most cells the targeting vector inserts randomly into the ES genome. However, in a few cells the targeting vector pairs with the cognate chromosomal DNA sequence and transfers the mutation to the genome by homologous recombination. Screening or enrichment procedures (or both) are then used to identify the rare ES cell in which the targeted event has occurred. The appropriate cell is then cloned and maintained as a pure population. Next, the altered ES cells are injected into the blastocoele cavity of a preimplantation embryo and the blastocyst is surgically transferred into the uterus of a foster mother, where development is allowed to progress to term. The resulting animal is chimeric in that it is composed of cells derived from both the donor stem cells and the host blastocyst. Genomic screening of these progeny is used to determine which mice received the allele carrying the targeted mutation. Interbreeding of heterozygous siblings yields animals homozygous for the desired mutation.

The power of gene targeting resides in the ability of the experimenter to choose precisely both the gene to be modified and the specific change to be introduced. Thus transgenesis and gene targeting are often directed towards different goals. The former is used to introduce new functions, the latter to augment or eliminate existing functions, e.g. fat content of meat.

TRANSGENESIS IN MAN?

Ethics apart, are transgenic humans a possibility in the near future? The answer must be in the negative for three basic reasons. First, microinjection has a high

Fig. 11.5 Gene targeting in the mouse (see text for details). In the example shown the ES cells are derived from a mouse homozygous for the black coat colour allele and the recipient blastocyst from an albino mouse. The fur of the chimeric mouse has patches of both colours because the mouse contains cells of both genotypes. Figure reproduced courtesy of Dr M.R. Capecchi and the editor of *Science*.

failure rate even in experienced hands and this presupposes that sufficient human fertilized eggs could be obtained. Second, microinjection of eggs can produce deleterious effects because there is no control over where the injected DNA will integrate in the genome. A high level of spontaneous abortion may be tolerated in farm animals but it would not be acceptable in humans. Finally, there is the question of limited usefulness. Most of the serious genetic disorders result in death before puberty or infertility in homozygous patients. Therefore the worst case that can be imagined is that both parents are heterozygous and, on average, only one-quarter of the offspring will suffer the conse-

quences of homozygosity. The question is, which ova are affected? Currently there is no way to tell and manipulating all ova poses far greater hazards!

GENE THERAPY IN HUMANS [13, 14, 15, 16]

As indicated earlier, only somatic cell gene therapy is being considered at present. This means that cells need to be extracted from the body, grown in culture, genetically manipulated and then reimplanted into the patient from whom the original tissue was taken. With current technology, this means that gene transfer in humans will be restricted to bone marrow and skin cells since only these cells can be maintained in culture. In turn, this severely restricts the number of genetic diseases for which therapy can be contemplated (Table 11.4). Studies with bone marrow cells are considerably more advanced than those with skin cells and only the former are considered here.

Bone marrow consists of a heterogeneous population of cells, most of which are committed to differentiation into erythrocytes, lymphocytes, etc. (see Box 11.1). Less than 1% of nucleated bone marrow cells are the undifferentiated progenitor cells known as stem cells. In gene therapy it is the stem cells which would be the primary target for exogenous DNA. Since they are low in number and not easily recognizable, an efficient gene delivery system is needed. With bone marrow cells calcium phosphate-mediated DNA uptake is very ineffective: the transfection frequency is as low as 1 cell in 10^6-10^7. Approximately 10^{10} cells can be obtained from the marrow of a patient undergoing transplantation and only 10^7-10^8 of them will be stem cells. Thus, at best, only 100 stem cells would be transfected by the calcium phosphate method and these would have little effect when transferred back into the marrow of the patient unless they had some extraordinary growth advantage.

Table 11.4 Target diseases for human gene therapy.

Protein	Disease state in absence of protein
Factor VIII	Haemophilia
Hypoxanthine-guanine phosphoribosyl transferase	Lesch−Nyhan disease
Pyrimidine nucleoside phosphorylase	Immunodeficiency
Adenosine deaminase	Combined immunodeficiency disease

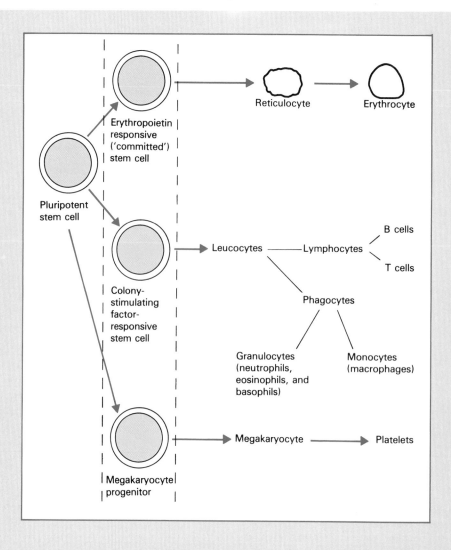

Box 11·1 The origin of the different blood cells

There are three formed elements in blood: erythrocytes (red blood cells), platelets and leucocytes (white blood cells). These elements are formed in the bone marrow from pluripotent undifferentiated cells known as *stem* cells. In response to erythropoietin, a hormone produced by the kidney, a proportion of stem cells undergoes maturation to produce the anucleate *reticulocyte*. Reticulocytes enter the circulation and mature into erythrocytes.

In response to a hormone, colony-stimulating factor, stem cells mature into five different kinds of leucocyte. Four of these different leucocytes are *phagocytic*, i.e. capable of digesting foreign debris in the bloodstream. The fifth class of leucocyte is the lymphocyte. There are two subsets of lymphocyte, the *B-cells*, which produce antibodies, and *T-cells*, responsible for the elimination of antibody-bound antigens.

Platelets are not cells but membrane-bound fragments of the cytoplasm of large cells known as megakaryocytes, which are found only in bone marrow.

Retrovirus vectors (see p. 124) offer a number of advantages over calcium phosphate-mediated DNA uptake. First, up to 100% of recipient cells can be infected and express the cloned gene. Second, as many cells as desired can be infected. Third, under appropriate conditions the DNA can integrate as a single gene copy, albeit randomly. Finally, the infection with the retroviral vector does not appear to harm cells.

In 1980 a haematologist tried, without success, to cure two patients of β-thalassaemia by adding viral genes to their bone marrow. That experiment was undertaken without appropriate approval and led to censure of the doctor concerned. In September 1990 the first fully approved experiment on human gene therapy was undertaken when doctors infused a 4-year-old girl with about 1 billion of her own white blood cells into which they had spliced copies of the gene that expresses adenosine deaminase (ADA). The girl's body lacks ADA and this results in severe combined immunodeficiency disease (SCID), which is usually fatal. ADA can be administered intravenously but it is degraded by the body almost immediately. Although it will be 12–24 months before the success of the experiment is known, a number of other children with ADA deficiency will be treated in the near future.

Following the successful demonstration of gene therapy, by experiments like those described above, there will be a need for *site-specific* expression of genes *in vivo*. Already it has been shown that a recombinant gene can be expressed in vascular tissue by implanting genetically modified endothelial cells at specific sites on the arterial wall. However, this technique requires that syngeneic (genetically identical) cell lines be established before genetic modification. This would make adaptation to the treatment of human disease very cumbersome. Recently it has been shown that a recombinant gene can be efficiently expressed at a specific site *in vivo* by direct introduction of genetic material at the time of catheterization. Many other advances are expected in the next few years.

Further reading

GENERAL REFERENCES

Hennighausen L., Ruiz L. & Wall R. (1990) Transgenic animals — production of foreign proteins in milk. *Current Opinion in Biotechnology* **1**, 74–8.

Old R.W. & Primrose S.B. (1990) *Principles of Gene Manipulation*, 4th edn. Blackwell Scientific Publications, Oxford.

Toltoshev P. & Anderson W.F. (1990) Gene expression using retroviral vectors. *Current Opinion in Biotechnology* **1**, 55–61.

CITED REFERENCES

1 Hammer R.E., Palmiter R.D. & Brinster R.L. (1984) Partial correction of murine hereditary growth disorder by germ-line incorporation of a new gene. *Nature* **311**, 65–7.

2 Selden R.F., Skoskiewicz M.J., Howie K.B., Russel P.S. & Goodman H.M. (1986) Regulation of human insulin gene expression in transgenic mice. *Nature* **321**, 525–8.

3 Meade H., Gates L., Lacy E. & Lonberg N. (1990) Bovine alpha$_{S1}$-casein gene sequences direct high level expression of active human urokinase in mouse milk. *Bio/technology* **8**, 443–6.

4 Pavirani A., Skern T., LeMeur M., Lutz Y., Lathe R., Crystal R.G., Fuchs J.-P., Gerlinger P. & Courtney M. (1989) Recombinant proteins of therapeutic interest expressed by lymphoid cell lines derived from transgenic mice. *Bio/technology* **7**, 1049–54.

5 Greaves D.R., Fraser P., Vidal M.A., Hedges M.J., Ropers D., Luzzatto L. & Grosveld F. (1990) A transgenic mouse model for sickle cell disorder. *Nature* **343**, 183–5.

6 Mullins J.J., Peters J. & Ganten D. (1990) Fulminant hypertension in transgenic rats harbouring the mouse *Ren-2* gene. *Nature* **344**, 541–4.

7 Van Brunt J. (1988) Molecular farming: transgenic animals as bioreactors. *Bio/technology* **6**, 1149–54.

8 Hammer R.E., Pursel V.G., Rexroad C.E., Wall R.J., Bolt D.J., Ebert K.M., Palmiter R.D. & Brinster R.L. (1985) Production of transgenic rabbits, sheep and pigs by microinjection. *Nature* **315**, 680–3.

9 Pursel V.G., Pinkert C.A., Miller K.F., Bolt D.J., Campbell R.G., Palmiter R.D., Brinster R.L. & Hammer R.E. (1989) Genetic engineering of livestock. *Science* **244**, 1281–8.

10 Rogers G.E. (1990) Improvement of wool production through genetic engineering. *Trends in Biotechnology* **8**, 6–11.

11 Rees W.D., Flint H.J. & Fuller M.F. (1990) A molecular biological approach to reducing dietary amino acid needs. *Bio/technology* **8**, 629–33.

12 Skarnes W.C. (1990) Entrapment vectors: a new tool for mammalian genetics. *Bio/technology* **8**, 827–31.

13 Kantoff P.W., Freeman S.M. & Anderson W.F. (1988) Prospects for gene therapy for immunodeficiency diseases. *Annual Review of Immunology* **6**, 581–94.

14 Friedmann T. (1989) Progress towards human gene therapy. *Science* **244**, 1275–81.

15 Nabel E.G., Platz G. & Nabel G.J. (1990) Site-specific gene expression *in vivo* by direct gene transfer into the arterial wall. *Science* **249**, 1285–8.

16 Weatherall D.J. (1991) Gene therapy in perspective. *Nature* **349**, 275–76.

Part V
Plant Biotechnology

12 Plant Cell, Tissue and Organ Culture

The *in vitro* culture of cells, tissues and organs of plants is one of the growth areas of biotechnology because of its potential to generate improved crops and ornamental plants. Such improvements have two components: the generation of genetic variation and the selection, maintenance and propagation of desired variants. The cultural techniques of plant cell, tissue and organ culture together with recombinant DNA technology are ideally suited to effecting both. At this point two popular misconceptions need to be corrected. First, these newer techniques will not appreciably shorten the time taken to introduce new plant varieties to the marketplace. This is because any new variety, however created, needs to be extensively field tested. Second, plant cell, tissue and organ culture are not synonymous; they are quite distinct techniques although they have many features in common. To appreciate each it is necessary to have a basic knowledge of the structure and development of flowering plants.

PLANT GROWTH AND DEVELOPMENT

When a seed germinates, a seedling emerges and grows into an adult plant with its many structures (Fig. 12.1). This plant consists of a stem and a root, each with an apical growth region, and many branches. In the stem the branches originate as buds which give rise to either leaves or flowers. As with stems and roots each bud also has its own apical growth region. Each apical growth region consists of *meristematic* cells, which are undifferentiated cells that are the sources of all cell types in the primary tissues of a plant.

Plant development involves a succession of events necessary for formation of new tissues and these events involve different classes of regulation; for example, leaves, flowers and fruit exhibit *determinate* growth, that is, growth ceases once they reach a certain size and shape. By contrast, roots and stems are *indeterminate* in growth for the meristems that produce them can continue to proliferate indefinitely provided conditions are suitable.

As the seedling emerges from the seed primary growth of the root occurs by multiplication of meristematic cells in the root apex. Above the meristem the cells grow in length without multiplication. Higher up, the outermost layer of elongated cells (the *epidermis*) can develop into root hairs, whose function is the absorption of water and nutrients. As the root grows some of the root cells undergo differentiation into the phloem and the xylem. The *phloem* consists of cells with perforated ends whose function is the conduction of nutrient sap. The *xylem* consists of cells with reinforced walls whose function is water conduction. Between the phloem and the epidermis is a mass of undifferentiated cells known as the *cortex*. While the root undergoes its development so too does the shoot. Division of the meristematic cells of the shoot apex results in growth of the stem and some of these stem cells differentiate into developing leaves (*leaf primordia*). Between these primordia and the elongating stem are axillary buds, also with meristems, which will give rise to branches or flowers.

Division of the apical meristem produces vertical or *primary* growth; increased width or thickness comes from *secondary* growth. Secondary growth starts with the appearance between the phloem and the xylem of a layer of meristematic cells known as the *vascular cambium*. These cambium cells multiply laterally and differentiate into xylem and phloem. As the xylem increases in thickness the innermost part dies and becomes the *pith*. Between the cortex and the epidermis another layer of meristematic cells differentiates into the cork cambium whose growth produces the tough outer layer, which on a tree is known as *bark*.

PLANT TISSUE CULTURE

Tissue culture is the process whereby small pieces of living tissue (*explants*) are isolated from an organism and grown aseptically for indefinite periods on a nutrient medium. For successful plant tissue culture it is best to start with an explant rich in undetermined cells, e.g. those from the cortex or meristem, because such cells are capable of rapid proliferation. The usual

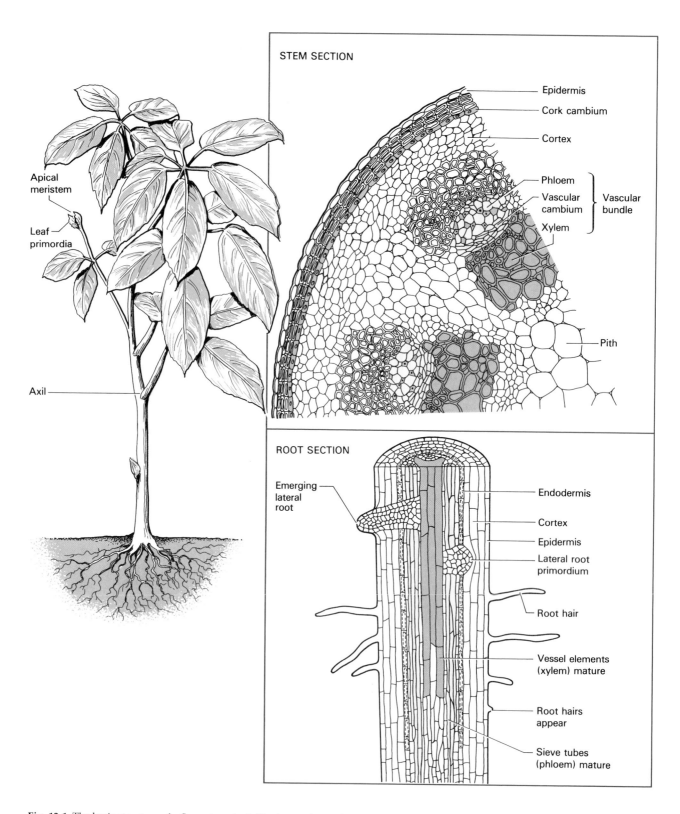

STEM SECTION

Epidermis
Cork cambium
Cortex
Phloem
Vascular cambium
Xylem
Vascular bundle
Pith

Apical meristem
Leaf primordia
Axil

ROOT SECTION

Emerging lateral root
Endodermis
Cortex
Epidermis
Lateral root primordium
Root hair
Vessel elements (xylem) mature
Root hairs appear
Sieve tubes (phloem) mature

Fig. 12.1 The basic structure of a flowering plant. The insets show a horizontal and a vertical section through part of a plant to illustrate the structural detail.

explants are buds, root tips, nodal segments or germinating seeds and these are placed on suitable culture media where they grow into an undifferentiated mass known as a *callus* (Fig. 12.2). Because the nutrient media used can support the growth of microorganisms the explant is first washed with a disinfectant such as sodium hypochlorite, hydrogen peroxide or mercuric chloride. Once established, the callus can be propagated indefinitely by subdivision (Fig. 12.3).

For plant cells to develop into a callus it is essential that the nutrient medium contain plant hormones, i.e. an auxin, a cytokinin and a gibberellin (Fig. 12.4). The absolute amounts of these which are required vary for different tissue explants from different parts of the same plant and for the same explant from different

Fig. 12.2 Close-up view of a callus culture.

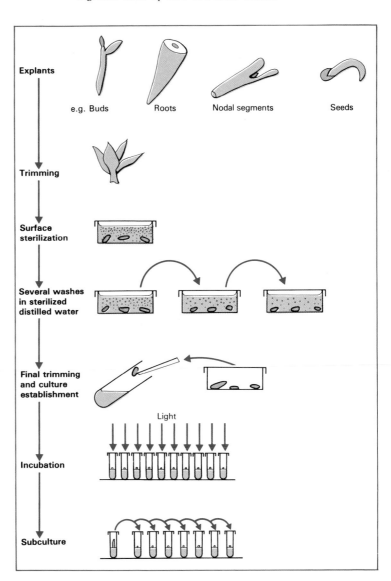

Fig. 12.3 Basic procedure for establishing and maintaining a culture of plant tissue. (After Mantell *et al.* 1985.)

A cytokinin
(N6-Methylaminopurine)

An auxin
(Indoleacetic acid (IAA))

A gibberellin

Fig. 12.4 The structures of some chemicals which are plant growth regulators.

Table 12.1 Composition of Murashige and Skoog (MS) culture medium.

Ingredient	Amount (mg/l)
Sucrose	30 000
$(NH_4) NO_3$	1 650
KNO_3	1 900
$CaCl_2.2H_2O$	440
$MgSO_4.7H_2O$	370
KH_2PO_4	170
$FeSO_4.7H_2O$	27.8
Na_2EDTA	37.3
$MnSO_4.4H_2O$	22.3
$ZnSO_4.7H_2O$	8.6
H_3BO_3	6.2
KI	0.83
$Na_2MoO_4.2H_2O$	0.25
$CoCl_2.6H_2O$	0.025
$CuSO_4.5H_2O$	0.025
Myo-inositol	100
Glycine	2.0
Kinetin (a cytokinin)	0.04–10.0
Indoleacetic acid	1.0–30.0

genera of plants. Thus there is no 'ideal' medium. Most of the media in common use consist of inorganic salts, trace metals, vitamins, organic nitrogen sources (glycine), inositol, sucrose and growth regulators. Organic nutrients such as casein hydrolysate or yeast extract and a gelling agent are optional extras. The composition of a typical plant growth medium is shown in Table 12.1.

PLANT CELL CULTURE [1]

When a callus is transferred to a liquid medium and agitated the cell mass breaks up to give a suspension of isolated cells, small clusters of cells and much larger aggregates (Fig. 12.5). Such suspensions can be maintained indefinitely by subculture but, by virtue of the presence of aggregates, are extremely heterogeneous. A high degree of genetic instability adds to this heterogeneity. Some plants such as *Nicotiana tabacum* (tobacco) and *Glycine max* (soybean) yield very friable calluses and cell lines obtained from these species are much more homogeneous and can be cultivated both batchwise and continuously. The cells can also be immobilized by the techniques described in Chapter 7.

Fig. 12.5 Photomicrograph of plant cells in suspension culture.

When placed in a suitable medium isolated single cells from suspension cultures are capable of division. As with animal cells (see p. 115), for proliferation to occur conditioned medium may be necessary. Conditioned medium is prepared by culturing high densities of cells of the same or different species in fresh medium for a few days and then removing the cells by filter sterilization. Media conditioned in this way contain essential amino acids such as glutamine and serine as well as growth regulators like cytokinins. Provided conditioned medium is used, single cells can be plated out on solid media in exactly the same way as microorganisms; instead of forming a colony as do microbes, plant cells proliferate to give a callus. In this way it is possible to isolate and propagate mutant cell lines. The occurrence of such mutants is known as somaclonal variation.

Protoplasts. Protoplasts are cells minus their cell walls. They are very useful materials for plant cell manipulations because under certain conditions those from similar and contrasting cell types can be fused to yield somatic hybrids, a process known as *protoplast fusion.* Protoplasts can be produced from suspension cultures, callus tissue or intact tissues, e.g. leaves, by mechanical disruption or, preferably, by treatment with cellulolytic and pectinolytic enzymes. Pectinase is necessary to break up cell aggregates into individual cells and cellulase to remove the cell wall proper. After enzyme treatment protoplast suspensions are collected by centrifugation, washed in medium without enzyme, and separated from intact cells and cell debris by flotation on a cushion of sucrose (Fig. 12.6). When plated on to nutrient medium protoplasts will in 5–10 days synthesize new cell walls and then initiate cell division.

PLANT ORGAN CULTURE [2, 3]

If apical shoot tips are surface-sterilized and placed in a growth medium lacking plant hormones, they will develop into single seedling-like shoots. If instead the medium is supplemented with a cytokinin, axillary shoots will emerge from their normal positions in the leaf axils and produce a shoot cluster. Once such clusters have developed they can be subdivided into smaller clumps of shoots or separate shoots, which will, in turn, form similar clusters when subcultured on fresh medium (Fig. 12.7). Provided the basic nutrient formulation is adequate for normal growth this subdivision process, which is known as *micropropagation,*

can be undertaken every 4–8 weeks for an indefinite period. Indefinite culture of plant roots can be achieved in a similar manner but in this instance the nutrient medium should contain an auxin and no cytokinin.

Many companies are using micropropagation to mass-produce ornamental plants. However, because the technique is time-consuming and labour-intensive, automated methods are being developed.

REGENERATION OF PLANTS [4, 5, 6]

Numerous species of plant can be maintained in cell or callus culture. Once in this state it is possible to isolate new genotypes by means of somaclonal variation, protoplast fusion or mutagenesis. However, such cell variants are of little value unless they can be induced to develop into intact plants. Cells which can differentiate into an intact plant are said to be *totipotent* and totipotency is a property of undetermined cells. Undetermined cells are, by definition, capable of switching to different pathways of development depending on the environment in which they are located. As might be expected, undetermined cells are those cells which most readily develop into a callus.

As indicated earlier, the formation and maintenance of callus cultures requires the presence of a cytokinin and an auxin, whereas only a cytokinin is required for shoot culture and only an auxin for root culture. Therefore it is no surprise that increasing the level of cytokinin to a callus induces shoot formation and increasing the auxin level promotes root formation. Ultimately plantlets arise through development of adventitious roots on the shoot buds formed, or through development of shoot buds from tissues formed by proliferation at the base of rootlets (Fig. 12.8). The formation of roots and shoots on callus tissue is known as *organogenesis.* The cultural conditions required to achieve organogenesis vary from species to species, and have not been determined for every type of callus. Historically it has proved to be difficult to regenerate plants from protoplasts of cereal and grass species. Thus recent successes with wheat and rice are highly commendable since they are among the most important crops in the world.

Under certain cultural conditions calluses can be induced to undergo a different development process known as somatic embryogenesis. In this process the callus cells undergo a pattern of differentiation, similar to that seen in zygotes after fertilization, to produce *embryoids.* Such cells are embryo-like but differ from

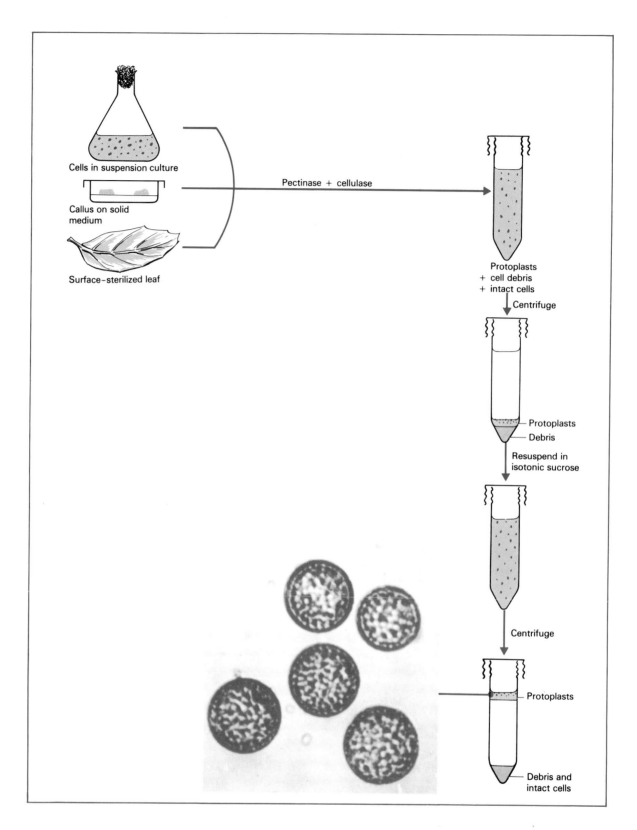

Fig. 12.6 Schematic outline of the enzymatic procedure used to isolate plant protoplasts. The inset shows a photomicrograph of protoplasts.

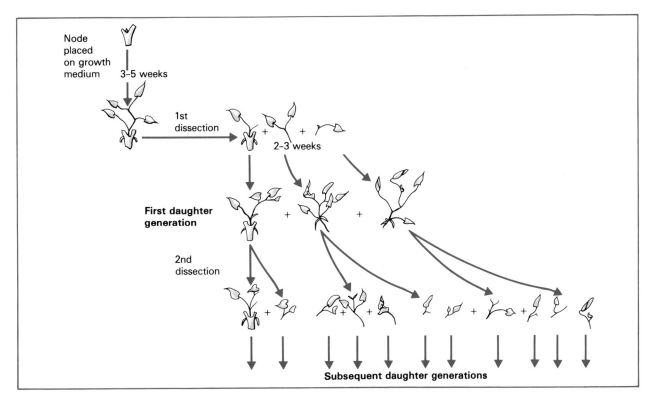

Fig. 12.7 Plant organ culture by means of axillary bud proliferation. (Reproduced courtesy of Dr S.H. Mantell.)

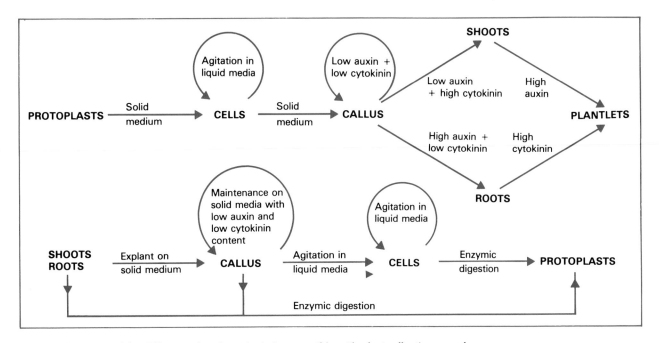

Fig. 12.8 Summary of the different cultural manipulations possible with plant cells, tissues and organs.

normal embryos in being produced from *somatic* cells and not from the fusion of two germ cells. These embryoids can develop into fully functional plants without the need to induce root and shoot formation on artificial media. The embryogenic response leading to embryoid formation is stimulated when calluses which were established in medium in which 2,4-dichlorophenoxyacetic acid (2,4-D) was the auxin are transferred to 2,4-D-free medium containing reduced sources of nitrogen, e.g. ammonium salts.

ANTHER AND POLLEN CULTURE (see Box 12.1)

Pollen grains are determinate cells whose normal development involves the formation of a pollen tube and male gametes. When placed on suitable nutrient media, e.g. MS medium, most pollen grains follow this development pathway but a few grains will form a callus instead. Instead of culturing isolated pollen grains it is possible to culture the intact anthers containing the developing pollen and this results in the formation of embryoids directly from pollen grains. These embryoids can be induced to develop into whole plants which are true haploids. (NB Seeds of most flowering plants are diploids.) As long as optimal conditions for donor plants and explants are provided, it is possible to obtain several hundred haploid plants *from a single anther*.

For successful anther and pollen culture it is essential to excise the flower buds at the correct time and usually this is at the time of the first mitotic division of the uninucleate microspore tetrads. Treatment of the excised buds or anthers prior to culturing on nutrient media can also increase the frequency at which pollen grains develop into plants and the treatment most often used is storage in the cold for several days to several weeks. Such cold stress probably raises the levels of growth regulators in the anthers. It should not be surprising, therefore, that inclusion of growth regulators in the culture medium can also be beneficial.

APPLICATION OF CELL, TISSUE AND ORGAN CULTURE

In this section only non-breeding applications of cell, tissue and organ culture will be discussed. The applications in plant breeding are discussed in Chapter 13.

Production of fine chemicals. For centuries plants have been an important source of precursors and products used in a variety of industries including those dealing with pharmaceuticals, food, cosmetics and agriculture (Table 12.2). Although in many cases natural plant products have been superseded by synthetic compounds, the plant kingdom is still a major contributor of speciality chemicals with a market value measured in billions of dollars. It is worth remembering that the more structurally complex the compound, and in

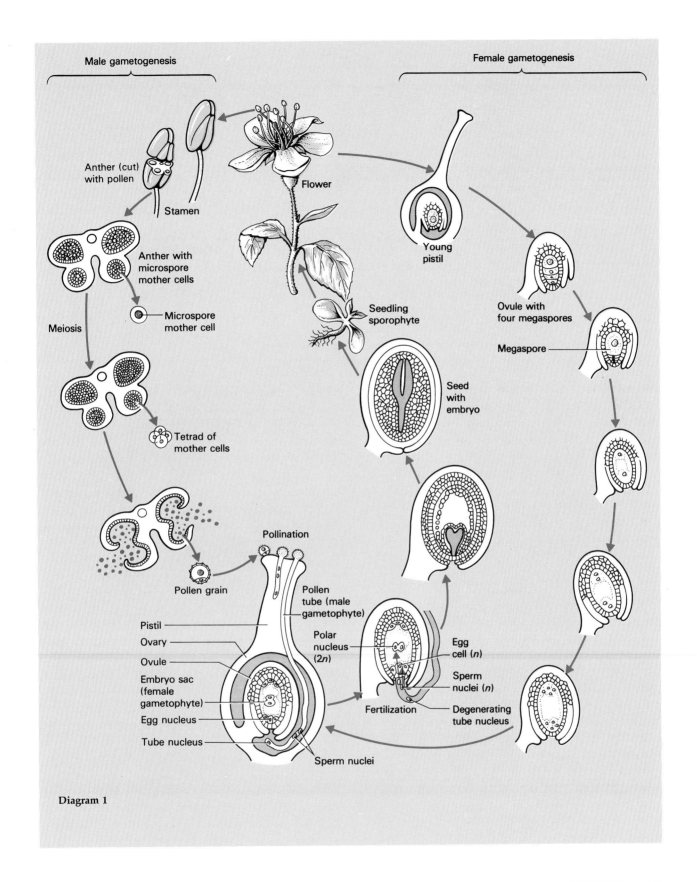

Male gametogenesis

Female gametogenesis

Anther (cut) with pollen

Stamen

Flower

Young pistil

Anther with microspore mother cells

Microspore mother cell

Ovule with four megaspores

Meiosis

Megaspore

Tetrad of mother cells

Seedling sporophyte

Seed with embryo

Pollination

Pollen tube (male gametophyte)

Pistil

Ovary

Ovule

Polar nucleus (2n)

Egg cell (n)

Embryo sac (female gametophyte)

Sperm nuclei (n)

Egg nucleus

Tube nucleus

Fertilization

Degenerating tube nucleus

Sperm nuclei

Pollen grain

Diagram 1

Product	Use	Source
Codeine	Analgesic	*Papaver somniferum*
Diosgenin Sitosterol	Raw material for production of pharmacologically active steroids	*Dioscorea deltoidia* *Zea mays*
Quinine	Antimalarial; embittering agent for food and drink	*Cinchona ledgeriana*
Digoxin	Treatment of cardiovascular disorders	*Digitalis* sp.
Scopolamine (hyoscine)	Treatment of nausea, especially motion sickness	*Datura stramonium*
Vincristine	Treatment of certain cancers	*Catharanthus roseus*
Atropine	Treatment of cardiac arrhythmias. Dilation of the pupil of the eye	*Atropa belladonna*
Reserpine	Treatment of hypertension	*Rauwolfia serpentina*
Pyrethrin	Insecticide	*Chrysanthemum* sp.
Jasmine	Perfumery	*Jasminum* sp.
Saffron	Food colourant and flavouring agent	*Crocus sativus*
Menthol	Flavouring	*Mentha piperita*

Table 12.2 Some chemicals obtained from plants.

particular the number of stereochemical centres in the molecule, the less likely it is that plant production will be replaced by chemical synthesis.

The demand for plant-derived chemicals is largely from the developed countries but they are produced in Third World countries, where political instability and lack of funds means that social disorder or plant disease could interrupt supplies at any time. Could plant cell culture provide an alternative source of such compounds? Given current technology an average process would have a yield of 20 g dry weight of cells per litre and a productivity of 10 g dry weight per litre per day. If 1% of the cell mass were product, then, for the process to be economically viable, the product would need to have a value of at least $500/kg. Nor is product value the only criterion; the size of the market is equally important. Not surprisingly, only one commercial process has emerged, the synthesis of shikonin by cells of *Lithospermum erythrorhizon*. Part of the problem is that many of those plant products which are of high value are not one but a mixture of compounds. Oils used as plant fragrances are a good example. Although all the various ingredients may be produced by culturing cells of the appropriate plant species, the proportion of the different components relative to one another will be wrong.

Where a plant product is a single chemical species, then improving the yield per unit cell mass would greatly improve the economics of cell culture. In this regard it should be noted that the yield of shikonin, the only plant cell culture product commercially available, is 14% of cell dry weight. One way of increasing product yield might be to mimic the methods used to increase penicillin production (see p. 86) and isolate high-yielding plant cell lines. Before genetic engineering techniques will be of much value it will be necessary to have a considerable amount of information about the biochemistry and genetics of product formation.

Over 1500 new compounds are identified in plants every year and it has been suggested that some of these could be developed into products produced in cell culture. The problem with this approach is that it is technology-driven and not market-led. By definition the compounds will be of high value and the most likely customer is the pharmaceutical industry. However, in recent years this industry has directed its drug-seeking activities at designing chemicals specifically to interact with key enzymes or biological effectors. The use of natural compounds is not fashionable.

Production of pathogen-free plants [7]. Crop and ornamental plants can be infected by a wide variety of microbial pests: viroids, viruses, mycoplasmas, bacteria, fungi and nematodes. Such infections greatly reduce the yield, vigour and quality of plants. For example, virus infection of ornamental plants reduces the size and number of blooms produced and infection of fruit crops reduces yield by up to 90%. In some cases infection can be eliminated by micropropagation of unaffected parts of the plants and this approach is particularly feasible if the problem is caused by bacteria, fungi or nematodes. Viroid, virus and mycoplasma infections are much more troublesome since they tend to be systemic. More specialized methods are required for their elimination and these methods are based on plant regeneration.

The basic method of obtaining virus-free plants is culture of apical meristems. If a small enough piece is taken and cultured, preferably less than the first half-millimetre of the growing tip, it is often possible to obtain virus-free material which can be proliferated *in vitro* and then used to regenerate virus-free plants. An alternative method is to take larger explants of apical meristem and subject them to heat treatment to reduce or eliminate virus replication. Suitable treatments are 30–37°C for 10–14 days or 50–60°C for 5–10 min. The third method is to incubate apical meristem explants on suitable culture media containing malachite green or thiouracil, which can reduce virus replication.

LARGE-SCALE PLANT PROPAGATION

If a plant can be propagated vegetatively, then it is a relatively easy matter to increase the number of plants available for cultivation. However, if a plant can only be propagated from seed, then increasing plant numbers can be a lengthy process. This is particularly true if one is dealing with shrubs and trees which may not flower until several years after seed germination. In such cases the use of *in vitro* methods of propagation can be invaluable. In practice, an explant may be taken and induced to form a callus on solid media and then a suspension culture derived by inoculating the callus into a liquid medium (Fig. 12.9). After the cells have increased in number they are plated out and somatic embryoids selected. On cultivation these embryoids will develop into regular plants. Large-scale clonal propagation of this kind is being used commercially with a number of tropical plantation crops, e.g. oil palms.

Another micropropagation method which is being used commercially is the production of minitubers. When axillary shoot cultures of potato are cultivated in the presence of appropriate levels of cytokinins and gibberellic acid they form large numbers of very small tubers (*minitubers*). These minitubers can be sown directly in the field and will generate normal plants.

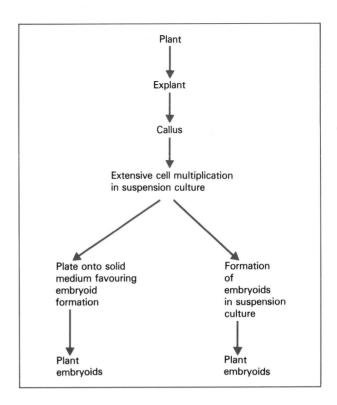

Fig. 12.9 Schematic outline of the method used for mass propagation of plants using *in vitro* culture techniques.

The advantages of this method of micropropagation can be seen from the data in Table 12.3.

PLANT GERMPLASM BANKS

In this century plant breeders have made enormous contributions to increased food production throughout the world. One important part of their work is the introduction of genetic diversity by intercrossing or mating selected germplasm with outstanding characters that complement one another. One source of this genetic diversity is from related species occurring in the wild, frequently in the less well-developed countries of the world. However, the numbers of wild species and their natural habitats are rapidly disappearing in these countries as a result of growing urbanization and an increase in the amount of land being cultivated. There is a general fear that potentially valuable germplasm is being lost irretrievably. As a consequence there is a growing use of gene banks. For many crops these comprise seed stores in which representative stocks of seed are held at a reduced temperature and humidity. The stocks are tested periodically

Table 12.3 Comparison of traditional and tissue culture methods for propagating tubers. The reduced number of plants obtained from the minitubers is an allowance for 5% loss on establishment in soil. (Reproduced from Mantell *et al.* 1985.)

Traditional	Micropropagation
Year 1	
100 g tuber	100 g tuber
↓	↓
One mature plant	One preconditioned plant
	↓
	10 nodal cultures
↓	↓
1600 g tubers	Shoot multiplication
↓	↓ *Dormant*
– – – – – – – – – – – – – – – – 65 000 minitubers – – – – –	*season*
Year 2	
16 mature plants	617 500 plants
↓	↓
16 × 600 g tubers	617 500 × 500 g tubers
= 25.6 kg	= 308 750 kg

for loss of viability and, if necessary, replenished by regrowth and harvesting.

There are two problems associated with seed banking. First, it cannot be used with so-called *recalcitrant* seeds. Such seeds are large and succulent, lack a dormancy phase and are incapable of surviving low temperatures and/or dehydration. Until now they have been maintained in small plantations set aside as *field gene banks*. Second, seed storage is not applicable for plants which are propagated vegetatively. Vegetative propagules such as bulbs and tubers can be stored but not for long periods of time.

As a result of recent advances, cell, tissue and organ culture is being considered as a means of germplasm storage. Two approaches are being adopted. One is *slow growth* involving the depression of metabolism by physical or chemical means and the other is the suppression of growth by storage at low temperatures (*cryopreservation*). Slow growth is only applicable to organized shoot cultures, whereas cryopreservation has proved most successful for cell cultures and shoot tips. Whatever method is used, it is essential that genetic variability should not occur during regeneration of the intact plant and this cannot be guaranteed.

Further reading

GENERAL REFERENCES

Glimelius K., Fahlesson J., Landgren M., Sjodin C. & Sundberg E. (1991) Gene transfer via somatic hybridization in plants. *Trends in Biotechnology* **9**, 24–30.

Mantell S.H., Matthews J.A. & McKee R.A. (1985) *Principles of Plant Biotechnology*. Blackwell Scientific Publications, Oxford.

Sen S.K. & Giles K.L. (1982) *Plant Cell Culture in Crop Improvement*. Plenum Press, New York.

Yeoman M.M. (1986) *Plant Cell Culture Technology*. Blackwell Scientific Publications, Oxford.

Details of all the techniques described in this chapter can be found in the six-volume treatise *Cell Culture and Somatic Cell Genetics of Plants*, Academic Press, New York. Volume I: *Laboratory Procedures and Their Applications*, ed. I.K. Vasil, 1984. Volume II: *Cell Growth, Nutrition, Cytodifferentiation, and Cryopreservation*, ed. I.K. Vasil, 1985. Volume III: *Plant Regeneration and Genetic Variability*, ed. I.K. Vasil, 1986. Volume IV: *Cell Culture in Phytochemistry*, ed. F. Constabel and I.K. Vasil, 1987. Volume V: *Phytochemicals in Plant Cell Cultures*, ed. F. Constabel and I.K. Vasil, 1988. Volume VI: *Molecular Biology of Plant Nuclear Genes*, ed. J. Schell and I.K. Vasil, 1989.

CITED REFERENCES

1 Larkin P.J. & Scowcroft W.R. (1981) Somaclonal variation — a novel source of variability from cell cultures for plant improvement. *Theoretical and Applied Genetics* **60**, 197–214.

2 Giles K.L. & Morgan W.M. (1987) Industrial scale plant micropropagation. *Trends in Biotechnology* **5**, 35–9.

3 Levin R., Gaba V., Tal B., Hirsch S., DeNola D. & Vasil I.K. (1988) Automated plant tissue culture for mass propagation. *Bio/technology* **6**, 1035–40.

4 Vasil I.K. (1988) Progress in the regeneration and genetic manipulation of cereal crops. *Bio/technology* **6**, 397–402.

5 Shinamoto K., Terada R., Izawa T. & Fujimoto H. (1989) Fertile transgenic rice plants regenerated from transformed protoplasts. *Nature* **338**, 274–6.

6 Vasil V., Redway F. & Vasil I.K. (1990) Regeneration of plants from embryogenic suspension culture protoplasts of wheat (*Triticum aestiuum* L). *Bio/technology* **8**, 429–34.

7 Wang P.J. & Hu C.Y. (1980) Regeneration of virus-free plants through *in vitro* culture. *Advances in Biochemical Engineering* **18**, 61–99.

13 Biotechnology and Plant Breeding

Conventional methods of plant breeding have been remarkably successful in generating improved varieties of many of the major world crops. Most of the improvement has been in yield and has occurred by the breeding and selection of lines resistant to pests and diseases, which can tolerate stressful environments and which are larger or are heavier croppers. There have also been improvements in the intrinsic quality of the crop, e.g. taste, protein content, etc. The long-range targets of biotechnology as applied to plant breeding are exactly the same; that is, the creation of new plant varieties with increased vigour and yield, the incorporation of value-added traits, and the development of disease-, herbicide- and pesticide-resistant crops.

Scientifically the most exciting developments in plant breeding are centred on the use of recombinant DNA to introduce new characteristics into plants and it is these developments which form the basis of part of this chapter. However, there are a number of non-recombinant approaches which are also worthy of discussion.

Biotechnology and plant breeding: non-recombinant approaches

Basically, there are two non-recombinant approaches which merit discussion. The first of these is the exploitation of spontaneous or induced variation in cultured plant cells or tissues. The second is the use of intra- and interspecific protoplast fusion to mediate genetic exchange.

SOMACLONAL VARIATION [1]

Somaclonal variation technology takes advantage of the naturally occurring genetic variation that appears in plants regenerated from somatic cells grown in tissue culture. This variation can pre-exist in the explant tissue but more usually it arises during the tissue culture procedure itself. The genetic events which occur include changes in nucleotide sequence, chromosome structure and chromosome number, and the phenotypic variability which results can also occur in the regenerated plant and be heritable. However, some of the phenotypic variability can result from the physiological response of the cells to the environment of the culture dish or vessel as well as from *epigenetic* changes. Epigenetic events reflect physiologically altered levels of gene expression that are relatively stable in that they persist through mitosis to be expressed by daughter cells. In contrast to altered phenotypes having a genetic basis, those resulting from epigenetic changes tend not to be expressed in regenerated plants or their progeny. Operationally, genetic changes are distinguished from epigenetic changes by their ability to be transmitted following sexual crossing.

The variability present in cell cultures is ultimately visible in populations of regenerated plants. Simple visual examination can be used to identify gross morphological changes such as plant height, growth habit, leaf shape and size, flower morphology and pigmentation. The frequency of visible variability is sufficiently high for most individuals to be altered in some way but not so high that the majority of those individuals possess deleterious alterations. However, most of the desired phenotypic traits are not associated with morphological changes and screening procedures have to be used. For example, early blight-resistant clones of a particular potato cultivar could only be identified by inoculating leaves of regenerated plants with toxin derived from *Alternaria solani*, the early blight fungus. In addition, where a crop is usually seed-propagated (e.g. cereals such as wheat, oats) rather than vegetatively propagated (e.g. potatoes) it is essential to show that the genetic change giving the desired trait has not caused infertility. Such infertility can arise as a result of chromosomal rearrangements and can only be detected by attempting sexual crosses.

Despite the many potential uses claimed for somaclonal variation (Table 13.1) the fact remains that thus far there is not a single example of any significantly important new variety of any major crop species de-

Table 13.1 Some useful plant variants obtained through somaclonal variation.

Plant	Improvement
Carrots	Improved 'snacking' characteristics (sweetness, crunchiness, crispness)
Celery	Improved 'snacking' characteristics
Potato	Resistance to late blight (*Phytophthora infestans*) and early blight (*Alternaria solani*)
Sugar cane	Resistance to eyespot disease (*Helminthosporium sacchari*), downy mildew and Fiji disease (leafhopper-transmitted virus). Increased sucrose content
Tomato	Increased solids content

veloped as a result of its use which is being cultivated commercially. When somaclonal variation is used in breeding, conventional screening is still required and with it the need for large amounts of land and labour. What would be particularly advantageous is a means of directly selecting the desired phenotypes. So far there has been some limited success in this area by the application of some of the methods developed for selecting mutants of microorganisms. Thus tobacco cells resistant to methionine sulphoximine and maize cells resistant to *Dreschslera maydis* toxin have been selected *in vitro* and plants regenerated from them have been resistant to wildfire disease and southern leaf corn blight, respectively. Mutagenesis can be used to increase the frequency with which such mutants occur. However, many traits of agronomic importance,

such as grain quality, pest resistance, etc., cannot be selected directly. Not only are these traits not expressed by cultured cells but the absence of an understanding of the molecular and cellular bases of these traits precludes the development of *in vitro* selection schemes. In the immediate future *in vitro* methods may be beneficial for the selection of plant mutants with improved amino acid content. As with microbial cells (see p. 82) this can be achieved by deregulating the biosynthetic pathways, and some examples are shown in Table 13.2.

PROTOPLAST FUSION [2, 3, 4]

Often it is desirable to cross two related species that are sexually incompatible. Sexual crosses as used in conventional plant breeding are clearly impossible but the same goal can be achieved by fusing the relevant plant protoplasts.

Plant protoplasts isolated according to the method described in Chapter 12 (see p. 157) can be induced to fuse with each other even when derived from different species. Two basic methods are used. Certain chemicals such as polyethylene glycol, dextran and poly-L-ornithine act as *fusogens* in that they promote protoplast fusion. Alternatively *electrofusion* can be used. In this procedure protoplast adhesion occurs in a non-uniform electrical field and fusion occurs when a short pulse of direct current is applied. When the two protoplasts first fuse a cell is produced which contains both nuclei and both cytoplasms. If the parent cells were not ident-

Plant	Mutation	Selection procedure
Maize	Resistance to southern corn leaf blight	Resistance to toxin produced by the causative agent (*Dreschslera maydis*)
Maize	Increased levels (×100) of threonine in amino acid pool of maize kernels. Total threonine content increased by 50%	Resistance to growth inhibition by a mixture of lysine and threonine
Tobacco	Resistance to wildfire disease (*Pseudomonas tabaci*)	Resistance to methionine sulphoximine, which is an analogue of the toxin produced by causative organism
Tobacco	Salt tolerance	Resistance to high levels of sodium chloride
Tobacco	Herbicide resistance	Resistance to herbicide incorporated in culture media

Table 13.2 Some useful plant mutants obtained following selection of mutant cells *in vitro*.

ical, then the product is a *heterokaryon*. If the two nuclei subsequently fuse, a mononuclear cell known as a *synkaryon* is produced. Although fusion also brings the plastids (chloroplasts and mitochondria) together, they do not stay together and eventually the plastids from only one plant predominate (Fig. 13.1).

After fusion the nuclear and cytoplasmic genomes reassort and recombine, resulting in a wide array of gene combinations not attainable through conventional breeding. The ability to transfer plastids is especially promising since traits such as cytoplasmic male sterility (see Box 13.1), disease resistance and herbicide resistance are encoded by the organelles and not the nuclear genome. The biggest disadvantage is that the scientist has little control over what genetic information is retained and what is eliminated. When

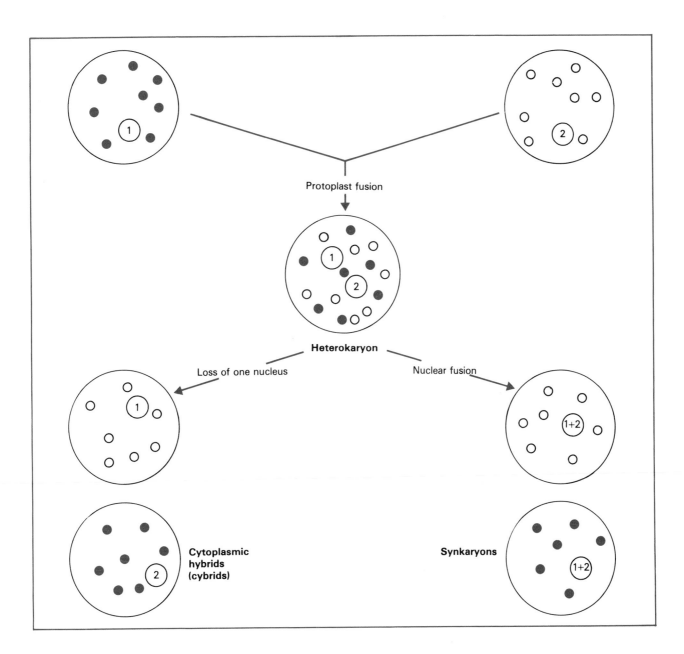

Fig. 13.1 The different kinds of hybrids which can be obtained following fusion of two different protoplasts. The nuclei of the two parental cells are labelled 1 and 2 and the plastids are shown as either red or black circles.

Box 13·1 Cytoplasmic male sterility

Many important crops such as maize, barley, wheat, sugar cane, sugar beet and onions are grown commercially from hybrid seeds. The importance of outbreeding to form hybrids, rather than inbreeding by self-fertilization, is that the hybrid progeny usually grow more vigorously and give higher yields of foodstuff. The production of hybrid seeds requires carefully controlled pollination. Maize, for example, is self-fertile and it is essential to block self-pollination if hybrid seed lines are to be maintained. Originally labour was employed to remove the tassels

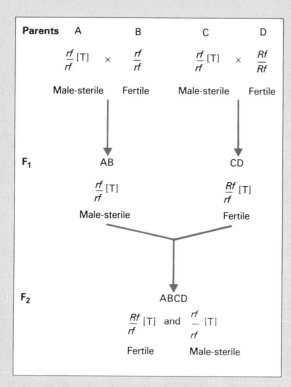

Diagram 1 The use of male-sterile plants in the production of hybrid maize. Note that only plants produced by outbreeding are shown. Plants can also be produced by inbreeding, i.e. B with B, D with D and CD with CD. However, inbred plants can be distinguished from outbred plants and the cobs not collected from them.

(pollen-producing part) of maize plants by hand. Today the cost of this would be prohibitive and male-sterile plants are used instead.

The maize seed which is sold commercially is the product of two sexual crosses involving four inbred lines represented in the diagram as A, B, C and D. Plants A and C carry a factor T (for Texas where it was discovered) which confers pollen sterility and these plants must be fertilized with pollen from another plant line, e.g. B or D. Pollen fertility can be restored by a dominant nuclear gene Rf. The F_1 hybrids both carry the factor T but, whereas the AB hybrids are male-sterile, the CD hybrids produce fertile pollen because they carry the gene Rf. In the F_2 generation the AB hybrid can only be pollinated by the CD hybrid although CD can fertilize itself. By careful choice of genetic markers it is possible to distinguish between the AB hybrids and the CD hybrids and only select corn cobs (source of seed) from the AB plants. Half of the seed in these cobs will be male-sterile and half will be fertile. When the seeds are sown there will be ample pollen from the fertile plants to pollinate the male sterile plants.

The cytoplasm of plant cells contains chloroplasts and mitochondria and these organelles possess DNA and their own protein-synthesizing machinery. There is now good evidence that the mitochondrial genome is involved in the phenomenon of male sterility; for example, restriction endonuclease analysis has shown differences between mitochondrial DNA from normal and male-sterile plants. Also, mitochondria from plants carrying the T factor synthesize a polypeptide of molecular weight 13 000 not found in normal plants. In the presence of the restorer gene, Rf, this polypeptide is not synthesized.

One disadvantage of male-sterile lines of corn is that they are particularly susceptible to southern corn leaf blight.

fusions are made between protoplasts from totally unrelated species, e.g. tobacco and soybean, the nuclei either fail to fuse or else one set of chromosomes is lost

following fusion. In contrast, fusions between sexually incompatible members of the same family have produced hybrid plants that retain some chromosomes

Table 13.3 Examples of somatic hybrids created by interspecific protoplast fusion.

Lycopersicon esculentum (tomato)	×	*Solanum tuberosum* (potato)
Datura innoxia (thorn apple)	×	*Atropa belladonna* (deadly nightshade)
Arabidopsis thaliana (thale cress)	×	*Brassica campestis* (turnip)
Petunia parodii (petunia)	×	*Petunia parviflora* (petunia)

from both parents (Table 13.3). The best known of these is the 'pomato' which is a potato–tomato somatic hybrid. Somatic hybrid plants such as the pomato are not of immediate value. Not only do many combinations prove infertile or fail to set seed but they need to be extensively back-crossed to eliminate unwanted portions of the alien genome. What they do provide is a means for introducing genes from unconventional sources. The most useful hybrids produced to date are cytoplasmic male-sterile rice, oil seed rape (*Brassica naponigra*) resistant to the fungus *Phoma lingam*, and *Solanum* (potato) varieties resistant to certain virus diseases.

ANCILLARY TECHNIQUES

Incompatibility, which is the failure of pollen to fertilize an ovule, is a major obstacle in conventional plant breeding programmes. There are a number of causes of incompatibility but common ones are failure of the pollen tube to develop and abortion of the young embryo. One solution to this problem is to fertilize the ovule *in vitro*. Pistils are cultured on nutrient media and, after removing part of the ovary wall, fertilization is effected by placing the pollen directly on the exposed ovule. Another cause of incompatibility is failure of the embryo to develop because nutrient starvation prevents the endosperm from developing. In this instance plants can be recovered for breeding purposes by culturing the embryo on nutrient media.

An essential requirement in many plant breeding programmes is plant stocks homozygous for genes of interest. Such stocks breed true but are laborious to prepare. Consider the cross AA BB cc dd × aa bb CC DD. All the progeny will have the genotype Aa Bb Cc Dd, i.e. will be heterozygous for all four loci. In order to obtain a pure-breeding stock of genotype AA BB CC DD up to five generations of back-crossing may be necessary. A much simpler alternative is to use anther or pollen culture. As has been discussed (see p. 160), plants regenerated from anther and pollen cultures are haploid but it is a simple matter to double the chromosome complement with colchicine treatment.

Colchicine interferes with mitosis and effects chromosome doubling by preventing sister chromatids from separating at anaphase.

IDENTIFYING QUANTITATIVE TRAITS IN PLANTS [5, 6, 7]

Many of the agronomically important traits in plants, such as yield, grain-moisture content, etc., have no obvious phenotype. This makes selection and measurement difficult in breeding programmes. Restriction fragment length polymorphisms (RFLPs), which were described in Chapter 3 (p. 34), may be of use here.

Example. Crop water use efficiency (WUE) is a measure of drought resistance. WUE can be correlated with the ratio of carbon-13 to carbon-12 in plants but this is not particularly useful in large-scale breeding programmes. However, the ratio can be predicted from three RFLPs which are additive. RFLPs associated with the solids content of tomatoes have also been identified.

RECOMBINANT APPROACHES TO PLANT BREEDING [8, 9]

Given the success of conventional plant breeding and the potential benefits of such biotechnological approaches as somaclonal variation and protoplast fusion, why bother to use recombinant methods? Two answers can be given to this question. First, genetic engineering can be used to introduce genes into a plant which do not exist in any member of the same plant family, or even in any plant. Second, gene manipulation in the guise of protein engineering could be used to make substantial alterations to key cellular proteins, changes which are unlikely to occur spontaneously.

Cloning vectors are required for genetic engineering of plants and two different kinds are available: those based on plant viruses and those derived from the Ti plasmid of the plant pathogen *Agrobacterium tumefaciens*. In addition, there must be a means of introducing the recombinant DNA into plant cells. Plant protoplasts can be induced to take up naked nucleic acid and the methods used are essentially no different from those used with animal cells (see Table 11.1, p. 142). However, handling protoplasts is technically tricky An alternative is to deliver nucleic acids directly into intact cells, using high-velocity microprojectiles (Fig. 13.2). A gunpowder charge fires spherical tungsten particles (average diameter 4 μm) carrying RNA or DNA, which are literally shot into the cell. This remarkable technique does not require cell culture or pretreatment of recipient tissue and circumvents the problems associated with regeneration of protoplasts. It is of particular value with cereals, which, in addition to not regenerating from protoplasts, exhibit limited susceptibility to *Agrobacterium* infection.

If genetically engineered plants are to be used commercially, then the following requirements must be satisfied:

1 introduction of the gene(s) of interest to all plant cells;

2 stable maintenance of the new genetic information;

3 transmission of the new gene to subsequent generations;

4 expression of the cloned genes in the correct cells at the correct time.

The extent to which these conditions are met by the two different vector systems is outlined below.

PLANT VIRUSES AS VECTORS

At first sight plant viruses appear very attractive as cloning vectors because not only can they infect intact plants but nucleic acid purified from them can also be infectious. However, not just any plant virus can be used for there are a number of other criteria which need to be satisfied. First, the virus must be able to spread from cell to cell via the plasmodesmata. Second, introduction of foreign nucleic acid into the virus genome might make it too large to be packaged in viral particles; thus the viral nucleic acid must be able to replicate and spread from cell to cell in the absence of viral coat protein. Third, the modified viral genome should elicit little or no disease symptoms in infected plants. Fourth, it would be preferable if the virus had a broad host range. Finally, since most genetic engineering centres around the *in vitro* manipulation of DNA it is preferable that the virus vector has a DNA genome. Only two groups of plant viruses, the caulimoviruses and the geminiviruses, have DNA genomes.

Caulimoviruses as vectors. Caulimoviruses are a group of spherical viruses (Fig. 13.3) which contain a circular double-stranded DNA genome. The archetypal

Fig. 13.3 Electronmicrograph of Cauliflower Mosaic Virus (CaMV). (Courtesy of M. Webb, National Vegetable Research Station).

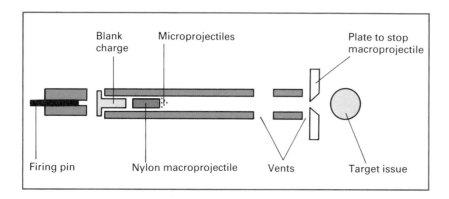

Fig. 13.2 Gun for delivering microprojectiles, coated with DNA, to plant tissue.

member is cauliflower mosaic virus (CaMV), hence the name of the group. The caulimoviruses are widely distributed throughout the temperate regions of the world and are responsible for a number of economically important diseases of cultivated crops. However, they have a restricted host range and are confined to a few closely related plants in nature.

One feature of CaMV which makes it attractive as a vector is that infection becomes systemic and very high levels of the virus are found in infected cells. In order for CaMV to be transmitted through the vasculature of the plant the DNA must be encapsidated in viral protein. Thus insertion of foreign DNA into the viral genome must not interfere with virus assembly. Unfortunately the CaMV genome has virtually no non-coding regions (Fig. 13.4) in which to insert foreign DNA although two regions of the genome, genes II and VII, do not seem essential for infection. The feasibility of replacing at least one of them with foreign DNA has been demonstrated. Gene II was replaced with a bacterial gene encoding methotrexate resistance and the chimeric molecule used to infect turnip plants. Not only did the turnips become systemically infected

but they also developed resistance to methotrexate applied as a spray.

The major limitation with CaMV vectors is capacity. If more than a few hundred nucleotides are inserted then infectivity is destroyed. Nor can helper viruses be used because homologous recombination between the helper and recombinant virus DNAs readily expels the foreign DNA with production of wild-type virus.

Geminiviruses as vectors. Geminiviruses derive their name from the fact that they have a geminate (paired particle) morphology (cf. Gemini, the heavenly twins). They have a very small, single-stranded circular DNA genome but replication is via a double-stranded intermediate. An attractive feature of the group is that they infect a wide range of crop plants which includes both monocotyledons and dicotyledons. However, purified viral DNA has never been successfully introduced into plants except with the help of an *Agrobacterium* vector.

RNA plant viruses as vectors. RNA-containing plant viruses have not been given much serious consideration as vectors because of the difficulty of joining

Fig. 13.4 Genetic map of the cauliflower mosaic virus genome. The eight coding regions are shown by the coloured boxes. The thin lines in the centre indicate the viral DNA strands, which are unusual in having discontinuities. IR1 and IR2 are two intergenic regions.

RNA molecules *in vitro*. This problem can be overcome by making cDNA copies of the RNA prior to *in vitro* manipulation. Indeed, a sequence encoding chloramphenicol resistance (chloramphenicol acetyltransferase) has been engineered into the brome mosaic virus genome in this way, although so far expression has been limited to plant protoplasts. This approach using a cDNA intermediate is slightly cumbersome but does have much to commend it. In particular, most plant viruses have genomes of RNA rather than DNA and thus the range of potential vectors is immediately expanded. Also, many of the RNA viruses have a filamentous (rod-shaped) morphology. The significance of this is that the size of the virus particle is determined by the length of the viral nucleic acid, i.e. there are no size restrictions on the nucleic acid to be packaged.

The disadvantage of viral vectors. As pointed out earlier, a key requirement for plant breeding is that the introduced genes be transmitted to subsequent generations. With virus vectors this is problematical. If the virus is not seed-transmitted, then the modified plants will need to be vegetatively propagated. Although this might be acceptable with certain crops, e.g. potatoes, there is always the possibility that the genetic information carried by the virus will be lost. It would be more satisfactory if the genes of interest could be introduced into a plant chromosome.

TI PLASMIDS AS VECTORS [10]

Crown gall is a plant tumour which can be induced in a wide variety of gymnosperms and dicotyledonous angiosperms by inoculation of wound sites with the bacterium *Agrobacterium tumefaciens*. Crown gall tissue represents true oncogenic transformation; callus tissue can be cultivated *in vitro* in the absence of the bacterium and yet retain its tumorous properties. These include the ability to form an overgrowth when grafted on to a healthy plant, the capacity for unlimited growth *in vitro* in the absence of plant hormones, and the synthesis of *opines*, which are unusual amino acids not found in normal tissue. The most common of these opines are octopine and nopaline (Fig. 13.5), although agropine or the agrocinopine family of opines may also be present. The type of opine produced is determined by the *Agrobacterium* strain causing the infection and not by the host plant.

Investigation of the molecular biology of crown gall disease has revealed that *A. tumefaciens* has evolved a

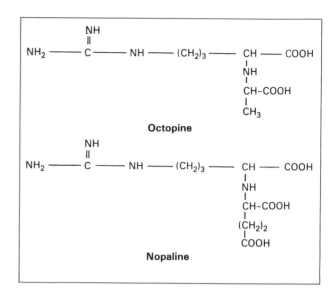

Fig. 13.5 The structure of the unusual amino acids octopine and nopaline.

natural system for genetically engineering plant cells. In particular, the oncogenic properties of *A. tumefaciens* reside in a plasmid known as the Ti(tumour-inducing)-plasmid. When the bacterium infects plant cells a portion of this DNA, called T-DNA, is transferred to the plant genome and it is the presence of the T-DNA which renders the cell oncogenic. The processes of DNA transfer and tumour formation are distinct (Fig. 13.6). The *vir* region of the Ti-plasmid encodes diffusible products which facilitate DNA transfer to the plant. The *onc* genes of the T-DNA itself function within the plant cell to produce and maintain the oncogenic phenotype. A number of loci have been identified, including 'large' (*tml*), 'shooty' (*tms*) and 'rooty' (*tmr*), which indicate the phenotype of the callus obtained when the loci are inactivated. The *onc* genes can be deleted from the T-DNA without affecting DNA transfer. However, for DNA transfer to occur the boundary sequences between the T-DNA and the rest of the Ti-plasmid must be intact.

Initial attempts to use the Ti-plasmids as plant vectors proved difficult because of the large size of the plasmids (mol. wt 200×10^6) and their genetic complexity. Consequently a system of intermediate vectors has been developed. Such vectors are bifunctional, being able to replicate in *Escherichia coli* and *A. tumefaciens*, and contain all or part of the T-DNA. The intermediate vector is manipulated *in vitro*, transformed into *E. coli* and then transferred to *A. tumefaciens*. In the latter

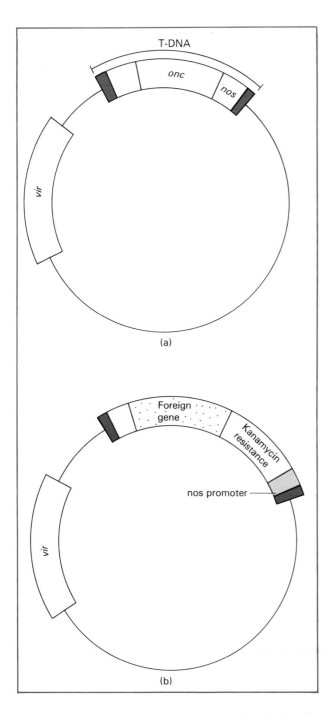

Fig. 13.6 (a) Schematic representation of a Ti-plasmid. *vir* is the virulence region whose presence is essential for T-DNA transfer to the plant genome. *nos* is the gene encoding nopaline synthase. *onc* is the group of genes whose gene products are responsible for tumour formation. The bold red areas are the DNA sequences essential for T-DNA transfer. (b) A modified vector for plant gene manipulation. The kanamycin-resistance gene under the control of the nopaline synthase promoter is essential for selection of transformed cells.

host *in vivo* recombination will occur and the T-DNA of the resident Ti-plasmid will be replaced with the mutant or genetically altered T-DNA. In this way a whole variety of foreign genes, e.g. interferon, ovalbumin and leghaemoglobin, have been transferred to tobacco cells.

For a foreign gene to be expressed in plant cells it is essential that it be preceded by a promoter recognized by the host cell. Since opine synthesis is not essential, vectors have been constructed in which the foreign gene is placed under the control of the *nos* (nopaline synthase) promoter. Thus transformed cells have been obtained which express antibiotic-resistance genes derived from bacteria. Ultimately the introduced genes need to be expressed in the correct tissue at the correct time. Progress is being made. When the antibiotic-resistance gene was placed under the control of the promoter from the gene for the small subunit of ribulose bisphosphate carboxylase, the level of antibiotic resistance observed was regulated by light as expected.

Another requirement for plant genetic engineering is the ability to transfer the cloned gene to every cell in the plant and for it to be passed to subsequent plant generations. This is not possible if genetically engineered *A. tumefaciens* is used directly to infect whole plants; the foreign gene will be restricted to the tumour. Furthermore, regeneration of plants from tumour tissue is difficult. The solution is to delete the *onc* genes from the T-DNA. In the absence of the *onc* genes transfer of T-DNA to plant cells is detected by selection for a cloned gene, e.g. antibiotic resistance. This technique has been further improved. Instead of infecting whole plants with *A. tumefaciens* the bacteria are co-incubated with leaf discs placed on tissue-culture medium supplemented with antibiotic. Those plant cells which become infected acquire antibiotic resistance and grow out of the disc as a callus. This callus is subsequently used to regenerate plants (Fig. 13.7).

The value of the Ti-system is demonstrated by the fact that not only is the chimeric gene present in all the cells of a plant regenerated from callus, but that such plants set seed and the chimeric gene is transmitted in that seed. The biggest drawback to the Ti-system is that in order to get transgenic plants one needs to start with cells which are *competent* for both regeneration and integrative transformation. Either alone is *not* sufficient. The most effective trigger for shifting cells to the competent state is enzymatic and/or mechanical wounding. Plant cells differ in their wound response and graminaceous plant species, especially cereals and

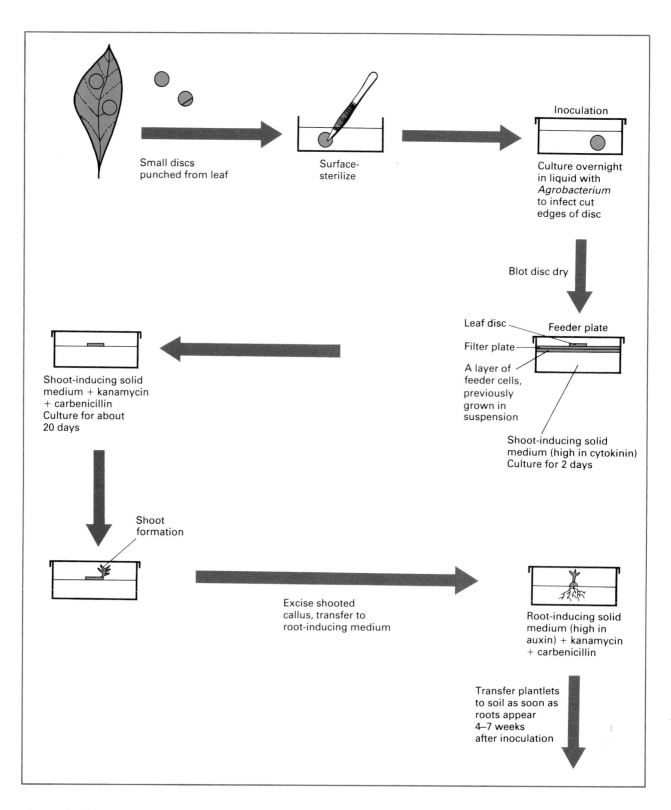

Small discs
punched from leaf

Surface-
sterilize

Inoculation

Culture overnight
in liquid with
Agrobacterium
to infect cut
edges of disc

Blot disc dry

Leaf disc
Filter plate
A layer of
feeder cells,
previously
grown in
suspension

Feeder plate

Shoot-inducing solid
medium (high in cytokinin)
Culture for 2 days

Shoot-inducing solid
medium + kanamycin
+ carbenicillin
Culture for about
20 days

Shoot
formation

Excise shooted
callus, transfer to
root-inducing medium

Root-inducing solid
medium (high in
auxin) + kanamycin
+ carbenicillin

Transfer plantlets
to soil as soon as
roots appear
4–7 weeks
after inoculation

Fig. 13.7 Leaf-disc transformation by Ti-plasmid vectors.

maize, which agronomically are the most important crops, have virtually no wounding response. As a consequence, *Agrobacterium*-mediated transgenesis of cereals has not been successful. Although maize streak virus DNA has been introduced to cereal cells via the Ti-system, the viral DNA replicates cytoplasmically and does not integrate into the host genome. Cells from those monocotyledonous plants which exhibit a wounding response, e.g. asparagus and yams, can be transformed with *Agrobacterium*, indicating that the Ti-system is not restricted to dicotyledonous plants.

Fig. 13.8 Effect of the herbicide Basta on plants with and without a Basta-resistance gene.

WHAT HAS BEEN ACHIEVED? [11]

Although the techniques for expressing heterologous genes in transgenic plants were described for the first time only in 1983 we are already witnessing the practical use of these methods for agriculturally relevant plant breeding. Not surprisingly, these examples have to do with the transfer and expression of single genes. Not only are single genes easier to manipulate but all gene insertion techniques for plants currently work with only one gene at a time.

Example: Herbicide resistance in plants [11]. Herbicides generally affect processes that are unique to plants, e.g. photosynthesis or amino acid biosynthesis. These processes are shared by both crops and weeds, and developing herbicides which are selective for weeds is very difficult. An alternative approach is to modify crop plants so that they become resistant to broad-spectrum herbicides. Two approaches to engineering herbicide resistance have been adopted. In the first of these the target molecule in the cell either is rendered insensitive or is overproduced. This has been achieved with 5-enolpyruvylshikimate-3-phosphate synthase, the target for glyphosate. A similar method has been used for engineering resistance to the herbicide Basta (Fig. 13.8). In the second approach a gene encoding an enzyme that degrades or detoxifies the herbicide is introduced into the plant. Thus resistance to gluphosinate and bromoxyinal was achieved by introducing bacterial genes encoding enzymes that inactivate those herbicides by acetylation or nitryl hydrolysis, respectively.

Example: Insect resistance in plants [11, 12, 13]. Insect-resistant plants are particularly desirable for a number of reasons. First, innate resistance eliminates the need for insecticides which are both costly and time-consuming to administer. Second, insecticides are not selective and can kill off desirable as well as undesirable insects. Finally, many insecticides accumulate in the environment and can lead to long-term changes in fauna. Transgenic plants have been produced which express the insect control toxin of *Bacillus thuringiensis* (see p. 76). When such plants were infected with caterpillars, damage to the leaves was minimal and all the larvae died within 3 days. Control plants not expressing the toxin gene were stripped bare. One problem with the use of *B. thuringiensis* toxins is that they are very specific for particular insects (see p. 77). An alternative approach has been to exploit the resistance to a wide variety of pests which is conferred on cowpeas by the presence of trypsin inhibitors. A gene for cowpea trypsin inhibitor has been cloned and expressed in tobacco plants, which then exhibited enhanced resistance to caterpillars.

Example: virus resistance [11, 14, 15, 16]. Major crop losses occur every year as a result of virus infections, e.g. tobacco mosaic virus causes losses of tomato plants of over $50 million per annum. Currently there is little that the farmer can do to control such infections and he is dependent on the plant breeder for the production of novel, virus-resistant cultivars. Unfortunately, little is known about the genetics of virus resistance. However, there is a useful phenomenon known as *cross-protection*, in which infection of a plant with one strain of virus protects against superinfection with a second, related strain. The mechanism of cross-protection is not understood but it is believed that the viral coat protein plays a vital role. Transgenic tomato, tobacco and potato plants have been generated which express the coat protein genes from one of the following viruses:

alfalfa mosaic virus, cucumber mosaic virus, tobacco mosaic virus, potato virus X and potato virus Y. In each case disease symptoms were greatly reduced in transgenic plants infected with the corresponding virus.

Example: Modification of food and feed proteins [17, 18, 19]. Thaumatins are a group of structurally related proteins which are produced by the West African plant *Thaumatococcus danielli* in the arils of the fruit. Thaumatins are 10 000 times sweeter than sucrose on a weight-for-weight basis and 10^{-8}M solutions taste sweet. At still lower concentrations, where sweetness is not detectable, thaumatins enhance many other kinds of flavour. A gene for one of the thaumatins (thaumatin II) has been introduced into potatoes using an *Agrobacterium* vector, and all the tissues of the transgenic plants tasted sweet—leaves, stems and roots, as well as the tubers. Since thaumatins are flavour enhancers as well as sweeteners, their introduction to fruits such as strawberries or melons could be particularly beneficial.

Increased demand for animal protein in human diets has caused a shift towards the intensive rearing of farm animals. This in turn has led to the use of soybean and corn meals as animal feeds. Such feed-stuffs are often deficient in sulphur-containing amino acids. One solution is to increase the methionine content of the seed proteins by introducing a heterologous gene encoding a methionine-rich protein. Suitable proteins exist in Brazil nuts, sunflowers, wheat and rice and the gene for one of them has been expressed in tobacco plants. Transgenic plants accumulated the methionine-rich protein up to 8% of total seed protein. It is expected that similar experiments will be successfully completed in other plant species in the near future.

Glycinin is one of the predominant storage proteins of soybean. To improve its functional properties, i.e. heat-induced gelation and emulsification, and its nutritional value, the proglycinin subunit has been modified. So far the modified proteins have only been expressed in bacteria but they did exhibit functional properties superior to those of the native glycinin. This establishes the possibility of creating theoretically designed novel glycinins with high food qualities.

Example: Transgenic plants as protein factories [20, 21, 22]. The overproduction of commercially valuable proteins in microbes and animal cells was described in earlier chapters. However, production of these proteins in plants might represent a cheaper alternative since no specialized facilities are required for cultivation. A number of therapeutic proteins such as enkephalins, insulin and human serum albumin have been produced in this way. More exciting perhaps is the production of antibodies. Transgenic tobacco plants producing either heavy or light immunoglobulin chains (see p. 139) were crossed to yield progeny in which both chains were expressed simultaneously. A functional antibody accumulated to 1.3% of total leaf protein and the antigen-binding characteristics were similar to those of a hybridoma-derived antibody.

Example: Novel pigmentation in decorative plants [23]. In many plants, flower colour is determined by the production of anthocyanin pigments. In an attempt to overproduce flower pigments in petunias, a gene encoding chalcone synthase (CHS), a key enzyme in anthocyanin biosynthesis, was introduced. The result obtained was surprising. The introduction of the CHS gene to foreign sites in the petunia genome was found to *suppress* the expression of the endogenous CHS genes, resulting in pure white flowers and ones with novel pigmentation patterns. This phenomenon is known as *co-suppression* and is believed to be due to a homology-dependent interaction between the introduced gene and the endogenous gene sequences.

Example: Engineering male sterility in crops [24]. Because of the importance of male sterility (see p. 168) plant breeders have been trying to develop new systems for inducing pollen failure. Recently a novel genetic method has been described which promises to be applicable to many plant species. The method relies on a promoter (TA29) from anther-specific genes in tobacco and which is functional only in the tapetum at a particular stage of floral development. The tapetum is a single layer of cells surrounding the pollen chamber inside the anthers and is important for pollen development. When an RNase gene was coupled to the TA29 promoter and introduced into tobacco or oil seed rape plants, the resulting transgenic plants were completely normal except that they did not produce functional pollen.

Further reading

GENERAL REFERENCES

Gasser C.S. & Fraley R.T. (1989) Genetically engineering plants for crop improvement. *Science* **244**, 1293–9.

Knight P. (1989) Engineered fruit and vegetable crops. *Bio/technology* **7**, 1233–7.

Kramer M., Sheehy R.E. & Hiatt W.R. (1989) Progress towards the genetic engineering of tomato fruit softening. *Trends in Biotechnology* **7**, 191–4.

Old R.W. & Primrose S.B. (1990) *Principles of Gene Manipulation*, 4th ed. Blackwell Scientific Publications, Oxford.

Vasil I.K. (1990) The realities and challenges of plant biotechnology. *Bio/technology* **8**, 296–301.

CITED REFERENCES

1 Larkin P.J. & Scowcroft W.R. (1981) Somaclonal variation — a novel source of variability from cell cultures for plant improvement. *Theoretical and Applied Genetics* **60**, 197–214.

2 Kyozuka J., Kaneda T. & Shimamoto K. (1989) Production of cytoplasmic male sterile rice (*Oryza sativa* L) by cell fusion. *Bio/technology* **7**, 1171–4.

3 Sjodin C. & Glimelius K. (1989) *Brassica naponigra*, a somatic hybrid resistant to *Phoma lingam*. *Theoretical and Applied Genetics* **77**, 651–6.

4 Gibson R.W., Jones M.G.K. & Fish N. (1988) Resistance to potato leaf roll virus and potato virus Y in somatic hybrids between dihaploid *Solanum tuberosum* and *S. brevidens*. *Theroretical and Applied Genetics* **76**, 113–17.

5 Ratner M. (1990) Identifying qualitative traits in plants. *Bio/technology* **8**, 401–3.

6 Martin B., Nienhuis J., King G. & Schaefer A. (1989) Restriction fragment length polymorphisms associated with water use efficiency in tomato. *Science* **243**, 1725–8.

7 McCue K.F. & Hanson A.D. (1990) Drought and salt tolerance: towards understanding and application. *Trends in Biotechnology* **8**, 358–62.

8 Klein T.M., Wolf E.D., Wu R. & Sanford J.C. (1987) High-velocity micro-projectiles for delivering nucleic acids into living cells. *Nature* **327**, 70–3.

9 Fromm M.E., Morrish F., Armstrong C., Williams R., Thomas J. & Klein T.M. (1990) Inheritance and expression of chimeric genes in the progeny of transgenic maize plants. *Bio/technology* **8**, 833–9.

10 Potrykus I. (1990) Gene transfer to cereals: an assessment. *Bio/technology* **8**, 535–42.

11 Gasser C.S. & Fraley R.T. (1989) Genetically engineering plants for crop improvement. *Science* **244**, 1293–9.

12 Delannay X., La Vallee B.J., Proksch R.K., Fuchs R.L., Sims S.R., Greenplate J.T., Marrone P.G., Dodson R.B., Augustine J.J., Layton J.G. & Fischoff D.A. (1989) Field performance of transgenic tomato plants expressing the *Bacillus thuringiensis* var. *kurstaki* insect control protein. *Bio/technology* **7**, 1265–9.

13 Perlak F.J., Deaton R.W., Armstrong T.A., Fuchs R.L., Sims S.R., Greenplate J.T., & Fischoff D.A. (1990) Insect resistant cotton plants. *Bio/technology* **8**, 939–43.

14 Stark D.M. & Beachy R.N. (1989) Protection against potyvirus infection in transgenic plants: evidence for broad spectrum resistance. *Bio/technology* **7**, 1257–62.

15 Lawson C., Kaniewski W., Haley H., Rozman R., Newell C., Sanders P. & Tumer N.E. (1990) Engineering resistance to mixed virus infection in a commercial potato cultivar: resistance to potato virus X and potato virus Y in transgenic russet burbank. *Bio/technology* **8**, 127–34.

16 Kaniewski W., Lawson C., Sammons B., Haley L., Hart J., Delannay X. & Turner N.E. (1990) Field resistance of transgenic russet Burbank potato to effects of infection by potato virus X and potato virus Y. *Bio/technology* **8**, 750–4.

17 Witly M. (1990) Thaumatin II — a palatability protein. *Trends in Biotechnology* **8**, 113–16.

18 Altenbach S.B. & Simpson R.B. (1990) Manipulation of methionine-rich protein genes in seed plants. *Trends in Biotechnology* **8**, 156–60.

19 Kim C.-S., Kamiya S., Sato T., Utsumi S. & Kito M. (1990) Improvement of nutritional value and functional properties of soybean glycinin by protein engineering. *Protein Engineering* **3**, 725–31.

20 Vandekerchove J., Van Damme J., Van Lijsebettens M., Botterman J., De Block M., Vandewiele M., De Clercq A., Leemans J., Van Montagu M. & Krebbers E. (1989) Enkephalins produced in transgenic plants using modified 2S seed storage proteins. *Bio/technology* **7**, 929–32.

21 Sijmons P.C., Dekker B.M.M., Schrammeijer B., Verwoerd T.C., van den Elzen P.J.M. & Hoekema A. (1990) Production of correctly processed human serum albumin in transgenic plants. *Bio/technology* **8**, 217–21.

22 Hiatt A., Cafferkey R. & Bowdish K. (1989) Production of antibodies in transgenic plants. *Nature* **342**, 76–8.

23 Jorgensen R. (1990) Altered gene expression in plants due to *trans* interactions between homologous genes. *Trends in Biotechnology* **8**, 340–4.

24 Mariani C., De Beuckeleer M., Truettner J., Leemans J. & Goldberg R.B. (1990). Induction of male sterility in plants by a chimaeric ribonuclease gene. *Nature* **347**, 737–41.

Part VI
Social Aspects of Biotechnology

14 Legal, Social and Ethical Aspects of Biotechnology

Recent advances in biotechnology have created many public policy and legal issues. A number of the most relevant are discussed here: patents, deliberate release of recombinant organisms, ethical and legal aspects of gene therapy, legal implications of DNA fingerprinting, biological warfare, and the future of biotechnology in Third World countries.

Scientists often find the practice of industrial law confusing; they are used to constructing theories based on the (supposed) rational examination of facts. Lawyers also examine facts but they interpret them within the framework of existing law. Unlike new theories, new laws are not created instantly and final acceptance comes only after months or years of courtroom debate. Reaching a final decision can take years. Nowhere is this clearer than in the field of patent law. Even though a patent examiner permits a patent to be issued, his decision may be challenged in the courts of law by a third party. Alternatively, a third party may deliberately infringe a patent in the belief that the patentee would lose his case if he chose to defend his patent rights in a court of law. Both situations have arisen recently regarding patents granted in respect of interferons. In one instance two major companies agreed out of court not to proceed with litigation because of the high costs involved and the uncertainties as to who would win.

Biotechnology and patent law [1]

WHAT IS A PATENT?

Patents are rights granted in respect of inventions capable of industrial application. These rights are not automatic but require action to register them—the filing of a patent application. After filing, the application is scrutinized by a patent examiner. If the examiner is convinced that the subject of the patent is both *novel* and *non-obvious*, i.e. involves an inventive step, the patent will be issued. Embodied within the novelty element is the absence of *prior disclosure*. In return for the legal protection given to the invention, the patentee must provide a written description of the invention which is sufficiently clear and effective to enable a person of *ordinary skill in the art* concerned to practise the invention from its teaching, i.e. a patent must be *enabling*. As normally constructed, the descriptive matter of a patent would be presented under a series of typical headings like those shown in Table 14.1.

CATEGORIES OF PATENTABLE INVENTION

Three categories of patent can be recognized and examples of each can be found in the field of biotechnology. These categories are:
1 compositions of matter (products);
2 processes for manufacture;
3 methods of use.

Product patents are the most effective form of protection because proof that the claim is being infringed by a competitor is usually apparent from an inspection or an analysis of the competitor's product, i.e. there is no need to enquire into the method of manufacture. In

Table 14.1 Typical descriptive sections found in a patent.

Heading	Content
Field of the invention	The technological field and the broad nature of the invention
Background of the invention	The problem to be solved, and a summary of prior art in the field and the disadvantages thereof
Objects of the invention	The nature of the improvements that the invention seeks to provide
Summary of the invention	A distillation of the essential elements of the invention
Detailed description of the invention	A chronological sequence of experimental steps to enable someone skilled in the art to repeat the work and recreate the invention
Detailed claims	A list of the specific aspects of the invention which are to be protected

addition, this is the only category of patent which has recognition throughout the US, Japan and Western Europe. A new antibiotic would normally be covered by a product patent and it would be irrelevant whether this antibiotic were discovered by a standard screening procedure or by using gene manipulation to generate novel combinations of antibiotic biosynthetic genes (see p. 86). A second example would be a novel micro-organism that could degrade a previously recalcitrant molecule, e.g. dioxin or Agent Orange, and which could be used to clean up toxic waste spills. It should be noted that until 1980 it was not possible to patent a living organism. However, in that year the US Supreme Court ruled (*Diamond* v. *Chakrabarty*) that the fundamental distinction between living organisms and inert matter was irrelevant for patent purposes.

Many of the products we expect to produce by recombinant DNA technology are not new; rather they are natural products, such as hormones, produced in quantity by an inexpensive procedure. As those products are not new it might be thought that they are not patentable but this is not necessarily so, as the example of US patent 2 563 794 shows. This patent has a single claim to highly purified vitamin B_{12} of defined chemical characteristics. The defendant in an infringement suit held that the patentees had not discovered vitamin B_{12} but merely recovered it from existing material in 'more nearly pure' condition. As such it was not a newly created compound but only a purified form of the anti-pernicious anaemia factor known to be present in liver. This argument was dismissed by the court on the grounds that until the patent was filed nobody had produced a product of comparable purity.

In recent years, most developed countries have adopted the product claim as an appropriate form of protection for new substances. Before this, in most countries of the world it was necessary, even in the case of a new product, to claim the invention in process terms. This unsatisfactory situation, which still exists in Latin America and Eastern Europe, requires alternative methods of preparation to be devised and the use of multiple process claims to cover every practical possible pathway to the product. The genuine process invention is one where a truly novel route or method has been devised or where new conditions have been found for operating a process of generally known type in order to achieve an improved result; an example here would be the production of indigo in *Escherichia coli* strains carrying a cloned naphthalene dioxygenase gene (see p. 88). Incidentally, this is a perfect example of a discovery being non-obvious. Another example would be the improved substrate utilization by strains of *Methylophilus methylotrophus* carrying a cloned *gdh* gene (see p. 78) and which are grown for single-cell protein production.

Process patents are weak patents in that it is difficult to know when a competitor has infringed your patent. For this reason many companies choose not to patent new processes but to retain them as *trade secrets*. The danger here is that a competitor may discover the same process and patent it and could then sue you for infringement! In such a situation the legal outcome is not guaranteed. Some of the factors to be considered before making a decision regarding filing a patent or keeping a trade secret are set out in Fig. 14.1.

Where a substance is not new but has been newly discovered to possess a valuable property it sometimes can be protected by a method of use patent. Very often this covers the substance in admixture with other substances in a form in which it will be utilized to take advantage of the new property. Patents covering the formulation of therapeutic proteins produced by recombinant DNA technology are now being published, the usual claim being increased stability.

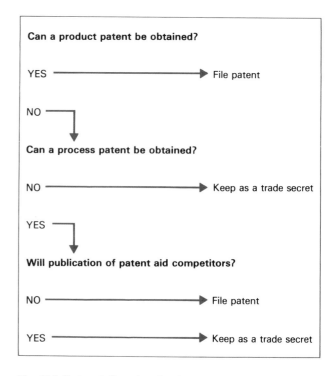

Fig. 14.1 Factors influencing the decision to patent or not to patent a product or process.

It was not until 1987 that the US Patent Office announced that life forms other than microorganisms were patentable. In 1988 it granted the first patent for a genetically engineered animal. This was a transgenic mouse which contains a gene that makes it extremely sensitive to cancer-producing agents. The mouse, which is sold commercially under the trademark Oncomouse, is used to test substances suspected of being carcinogenic.

In contrast with the decision of the US Patent Office, the European Patent Office (EPO) has rejected the application for a patent for the transgenic mouse. This was because Article 53(b) of the European Patent Convention prohibits the patenting of plant or animal varieties. Plant varieties were excluded because a decade earlier a system of protection, called Plant Variety Rights, was introduced. This gave more limited rights than a patent to the breeder of a new plant variety. Animal varieties were also excluded, although no legal protection for them existed then or now.

Both the US Patent Office and the EPO have announced that patenting humans is prohibited!

THE UNCERTAINTY OF PATENT LAW [2]

Patent litigation is an inexact science and infringement suits in biotechnology will often be reduced to debates between opposing expert witnesses. The situation is exacerbated by variations in patent law between different countries. For example, the requirement for absence of prior disclosure in the US means that the subject matter of the patent application should not have been described in public more than 1 year before the submission of the patent application. In Europe and the rest of the world patent applications must be filed before any publication. A better example, perhaps, concerns the *doctrine of equivalence*. Thus, a claim to the production of indigo in *E. coli* would not literally be infringed by an analogous process in another genus, e.g. *Pseudomonas*. However, an appeal against infringement might be upheld under the doctrine of equivalence. The courts generally allow a broader scope and wider range of equivalence to patents that represent a new field or a distinct step in the progress of an art. A countervailing doctrine, however, recognizes that the behaviour of chemical or biochemical compounds is not always predictable as conditions vary, and this should limit the scope of

claims to only those which a patent specifically teaches. Because of these uncertainties it is essential to employ a skilled legal draughtsman in the preparation of a patent; ideally he should be trained in a biological science and law.

A good example of the uncertainty of patent law is the narrow interpretation of the prohibition on patenting of plant varieties (see above) taken by the EPO. First, the prohibition does not apply to a plant modified other than genetically. Second, inventions relating to plants of one or more 'species' are potentially patentable. Finally, the prohibition does not apply to plants produced by a microbiological process. The EPO's interpretation of this term includes virtually all modern genetic engineering methods for creating plants. One patent granted recently concerns a technique to introduce a gene that produces phaseolin, a storage protein, into alfalfa, thereby increasing the crop's protein content and nutritional value. The patent covers the technique and any plants produced by the technique. Since it is not limited to plants of a particular variety, the EPO took the view that prohibition under article 53(b) did not apply. The EPO decision is currently being challenged by an unholy alliance of multinational pharmaceutical and chemical companies, animal rights groups and members of green political parties! Given the EPO view on patenting plants, it is surprising that it rejected the application for a patent for the Oncomouse.

Legal implications of DNA fingerprinting [3, 4]

Forensic biology is the testing of biological materials, the results of which may be used as evidence in legal and criminal proceedings associated with such diverse cases as paternity disputes, sexual assault and murder. In the last two decades there have been considerable technological advances as a result of the discovery of many new genetic polymorphisms among human proteins. The use of electrophoretic and immunological techniques to identify these new polymorphisms has worked its way into the forensic science repertoire with the result that more definitive genetic typing information can be provided for the legal system. However, these methods are not ideal. In the case of sexual assault the presence of spermatozoa indicates that sexual activity has occurred. The analysis of proteins from these spermatozoa, recovered intravaginally or as a dried strain, can be used to increase the probability of making a correct identification of the assailant but a thorough analysis is difficult. Only the ABO blood group and a select number of human leucocyte antigens (HLA) have been typed on sperm and the polymorphic red cell enzymes found in semen are limited in their phenotypic diversity. As a result, the likelihood of excluding a falsely accused individual is low; alternatively, there is a high probability that a random individual from the population might have the same markers as the assailant. Against this background what is the forensic potential for the DNA polymorphisms which can be detected with gene probes (see p. 38)?

The scientific conditions which must be satisfied before any blood group or other protein polymorphism can be used in a forensic science context are listed in Table 14.2. Since these protein polymorphisms have a genetic basis it is logical that DNA polymorphisms satisfy the same criteria. In the case of simple restriction fragment length polymorphisms (RFLP), like those shown in Fig. 3.8 (p. 36) for sickle-cell anaemia, there is no problem in fulfilling conditions 1−3. However, they are subject to the same criticisms as protein polymorphisms; that is, their lack of phenotypic diversity means that their diagnostic potential is limited. The diagnostic potential of the much more complex RFLP exhibited by minisatellite DNA (Fig. 3.11, p. 38) is far greater but it is much harder to satisfy the criteria laid down in Table 14.2. The solution is to test each sample with defined probes for four or five different loci. Each probe detects two alleles per locus and the combined probabilities of the alleles at the various loci give a rare combination pattern for each person. Even where different alleles at a particular locus have similar, but not identical, electrophoretic mobility, the remaining probes should provide sufficient differences.

Although the techniques involved in DNA fingerprinting are routine in molecular genetics laboratories, they are much too complex for any routine diagnostic laboratory. Indeed, as a number of well-documented cases in the USA have shown, even professional laboratories need to be exacting in their technique. A major technical issue is the phenomenon of band shifting. This problem occurs when DNA fragments migrate at different speeds through separate lanes on a single gel. It has been attributed to a number of factors including variability in the preparation of gels, the concentration of sample DNA, the amount of salts in the DNA solution and contamination. Band shifting has been a cornerstone of defence lawyers' and others' attempts to cast doubt on DNA profiling reliability. If mobility differences exist, they say, bands will not line up as they should. Judgements of 'matches' and 'non-matches' become too subjective to hold up in court. To overcome this problem, forensic DNA laboratories are developing probes that bind to monomorphic loci, restriction-enzyme fragments that are the same size in every person. Thus, if the monomorphic regions are displaced, one would know that band shifting had occurred and could apply a correction factor. Regardless of the technical problems which need to be resolved to prove guilt, it is worth noting that in 1988 and 1989 the US Federal Bureau of Investigation (FBI) *excluded* 144 individuals as suspects in over 400 crimes as a result of using DNA profiling.

A British chemical and pharmaceutical company, Imperial Chemical Industries (ICI), has set up a number of test centres in Europe and the US which provide the necessary expertise. There is a forensic problem associated with this because the prosecution and defence should have access to independent tests. There is considerable unease about there being only one centre to which samples can be sent. One solution in Britain would be for the Home Office Forensic Science Service to carry out 'fingerprinting' for the police, while the defence would use the ICI centre. Alternatively, separate scientists acting for the two sides could have independent access to the same facility, as sometimes occurs already in forensic science laboratories.

Table 14.2 Criteria for the use of any protein polymorphism in a forensic science context.

1 The marker used must be validated as a true genetic character. Its mode of inheritance must be demonstrated by family studies and it must be shown to be stable in an individual over a lifetime

2 Gene frequencies for the major population groups must be established to provide a base for the interpretation of findings

3 There must be a nomenclature to define the variants in each polymorphic system; this is necessary both for purposes of record keeping and for communication of data.

4 Standard types and typing reagents must be available to the forensic science community to allow independent testing in different laboratories.

5 Methodological guidelines need to be defined to ensure reliability of test results

6 Ultimately, markers must pass the barrier of blind trial testing. This establishes that the marker does not possess inherent ambiguities that might lead to typing error. The nature of the blind trials depends on whether the targeted use is paternity testing or criminal evidence.

Deliberate release of genetically engineered organisms [5, 6, 7, 8, 9]

GENETICALLY ENGINEERED MICROBES [10]

The very first commercial applications of recombinant DNA research have used genetically engineered microorganisms in fermentation technologies to make useful pharmaceutical products. Soon a host of other commercially useful products derived from gene manipulation will reach the market-place. While there has been some debate about the large-scale culture of recombinant organisms, particularly with regard to the safety of workers and the consequences of accidental release, concerns are abating rapidly. Partly this is due to the fact that modern fermentation plant can be made virtually leak-proof and the organisms killed at the end of the process; partly it stems from the fact that recombinant strains of E. coli and yeast, the current organisms of choice, do not have a propensity to colonize the gut. By exercising relatively simple precautions, companies can ensure that the risks posed by fermentation of well-characterized recombinant strains are not appreciably greater than the risks associated with traditional fermentations. Despite this, companies in West Germany had been prevented until recently from carrying out large-scale production of therapeutic proteins, e.g. insulin in E. coli, because *enabling* legislation was not in place. In contrast, in France the large-scale cultivation of recombinant E. coli or yeast does not require any authorization.

Genetically engineered microorganisms can also be commercially useful in ways that require that they be introduced on a large scale into the natural environment, i.e. *deliberate release*. Applications include pollution control, tertiary oil recovery and, most controversial of all, ice nucleation-defective bacteria for frost damage control (see Box 14.1). The feature which causes most concern is that the microorganisms may proliferate in the environment. Indeed, proliferation is often part of the *modus operandi* of the application. If, after release, we discover some unanticipated detrimental effects, there is no way we can destroy all the offending microbes.

Attempts have been made to introduce suicide genes into bacteria but invariably there is selection for those rare mutants which have lost the suicide gene. Furthermore, since we cannot see bacteria directly, monitoring their spread relies on culturing or hybridization. The polymerase chain reaction is a great asset in this

Box 14·1 Bacterial ice nucleation [5]

Frost-sensitive plants cannot tolerate ice crystal formation within their tissues. Ice crystals within sensitive plant tissues propagate rapidly, both intercellularly and intracellularly, causing mechanical breakdown of plant tissue and subsequent death. Many liquids, including water, do not invariably freeze at the melting-point of the solid phase. These liquids can be supercooled below the melting-point of the solid phase, e.g. water can be supercooled to −10°C to −20°C. The water−ice phase transition requires the presence of a catalyst or *ice nucleus*. Plants do not have intrinsic ice nuclei active at temperatures above −5°C but certain bacterial species can act as ice nuclei and thus have a primary role in limiting supercooling and inciting frost damage to frost-sensitive plants. The commonest ice nucleation-active bacteria isolated from plants are *Pseudomonas syringae* and *Erwinia herbicola*. They can initiate ice formation at temperatures of −1.5°C to −5°C, most probably by means of an outer membrane component. The gene(s) for ice nucleation have been cloned and are under active study.

Ice nucleation-active bacteria are present in large numbers in all temperate regions of the world and may be important in initiating rain and snow. Currently these bacteria are used to facilitate the formation of 'artificial' snow on ski slopes. Water containing *P. syringae* is sprayed through a fine nozzle on to a fan and the expansion-induced cooling produces snow.

respect because it permits the detection of one organism per gram of soil.

An aspect of deliberate release which seems to be forgotten is that for over 50 years microbiologists have been adding cultures of microorganisms (*inoculants*) to the environment in the hope of promoting a beneficial change, e.g. nodulation of legumes, oil and chemical waste removal, plant residue decomposition and destruction of plant pests. No substantial damage of any significance has been caused through such practices. There is no reason to think that a bacterium or fungus that is known not to damage the environment will cause environmental problems after it has obtained several well-characterized foreign genes. Although genetically engineered microbes may exchange their genetic information with indigenous soil organisms this should be of little concern; unrelated microbes have been sharing and reassorting their genes since time immemorial using the same methods as geneticists, e.g. transposon mutagenesis, phage-mediated transduction and plasmid transformation and conjugation. There are many plant pathogens that can naturally exchange genes with *E. coli* but we have yet to isolate a phytopathogenic *E. coli*.

Another question which must be asked concerns the ability of a newly introduced microorganism to predominate in a particular habitat already filled with a complexity of microbes which have evolved to compete successfully for all available niches. Concerns about the competitiveness of deliberately released micro-organisms stem from experience with the introduction of exotic weed species whose proliferation has occasionally been uncontrollable. However, in an area of natural habitat of, for example, 1 acre, there may be 10–50 different plant species; there could be many more different genera of microorganisms on a single plant. Regardless of whether the environmental dangers associated with deliberate release are real or imaginary it makes good sense to follow testing protocols before an organism is released. But how does one design relevant tests when environmental damage can only be assessed in field tests? For this reason there is a demand for an '*environmental impact*' statement before approval for deliberate release can be given.

Although environmental risk assessment is believed to be analogous to the human health risk assessment currently practised by institutional biosafety committees, nothing could be further from the truth: man is essentially a single habitat, the environment is not. The requirement for an environmental impact statement prior to deliberate release of a recombinant leads to bizarre situations. For example, the current controversy surrounding deliberate release stems from the wish of an investigator to attempt to stop frost damage to potato plants by spraying them with a genetically engineered strain of *P. syringae*. The strain in question is recombinant only in so far as gene manipulation was used to introduce a *deletion* in a gene for ice nucleation; this deletion renders the strain non-phytopathogenic. Whereas deliberate release of this organism has been subject to complex litigation, there is nothing to stop the investigator from deliberately releasing a mutant with similar proper-

ties created by conventional microbial genetics!

American regulators have finally decided that what is important is the end-product, not the method by which it was produced. Thus an ice nucleation-defective mutant poses the same risk regardless of whether it was produced by conventional genetic means or by recombinant DNA technology. Such a pragmatic view has not been adopted by the European Community. They argue that genetic engineering, i.e. the technique itself, poses new risks and so the products it produces should be regulated by new laws. Already there are 14 separate regulatory initiatives under consideration!

GENETICALLY ENGINEERED PLANTS AND ANIMALS [11, 12, 13]

There are fewer problems associated with the 'deliberate release' of genetically engineered plants and animals. For a start, since only large (crop) plants and domestic animals will be subject to gene manipulation, their spread in the environment can be observed. This contributes significantly to peace of mind. Since farm animals are normally kept in captivity, transgenic cows or pigs generate little concern. Transgenic fish could raise problems because the introduction of certain fish species as biological control agents has occasionally had disastrous consequences.

Some agronomists claim that the introduction of genetically engineered plants is analogous to the importation of exotic plant species and could disturb the existing ecological balance in an unpredictable way. But are not the major crops of the Western world, e.g. soybeans, wheat and rice in the US, foreign introductions and has their cultivation not been environmentally harmless? The consequences inherent in the release of genetically engineered crop varieties appear to differ little from those following the release of any conventionally bred new variety. There appear to be two concerns. The first arises because many commercial plant varieties are improved by crossing them with related weedy species. Therefore genetic engineering efforts to accomplish herbicide tolerance and increase photosynthetic efficiency could, as a result of introgression, result in fitter weeds. Such concerns seem misplaced if plant breeders are using the traditional methods to achieve the same ends. The second issue arises because the vectors used for genetic engineering of plants are derivatives of pathogenic microorganisms. If experiments are done with vectors which are not autonomous replicons, and if the newly introduced

DNA is stably integrated, genetically engineered plants should not serve as sources for disease transmission in the field.

Although it is highly likely that transgenic plants do not pose a threat to the environment, there is widespread public concern in many countries. This concern extends from the destruction of crops of transgenic plants in field trial plots to attempts to introduce legislation to ban field tests altogether. Attempts have been made to block field trials in West Germany of petunias with altered pigmentation and Dutch trials of potatoes resistant to the herbicide Basta. Both should be trialled by 1991. So should trials in Denmark, approved by parliament, of two new sugar beet varieties — one with a gene conferring resistance against 'rhizomania', a disease caused by beet necrotic yellow vein virus, and the other with a gene for resistance to the herbicide glyphosate. In contrast, over 50 separate releases had been made in France by the end of 1989, considerably more than in the rest of Europe. In the USA, a complex mixture of local, state and federal politics plus numerous lobby groups promises to cloud the issue for some time to come.

Despite the differences in opinion between the different protagonists in all countries, one issue surfaces repeatedly. Those opposed to the cultivation of transgenic plants feel strongly that the use of herbicide-resistant plants will lead to an increased use of herbicides, which are considered to be environmentally unfriendly. The companies sponsoring the use of such transgenic plants claim that the commercial strategy in engineering herbicide resistance is to gain market share through a shift in herbicide use. Thus, they say, herbicide-resistant plants will have the positive impact of reducing overall herbicide use through substitution of more effective and environmentally acceptable products!

RECOMBINANT ORGANISMS AND THE MANUFACTURE OF FOOD AND DRUGS [14, 15, 16, 19]

In the next few years it is anticipated that a whole series of new drugs, principally human proteins and viral vaccines, will be available to the physician as a result of developments in recombinant DNA technology. Should these drugs be regulated any differently from existing drugs produced by more conventional means? Within the drug regulatory agencies there is a growing recognition that this should not be the case.

The reason is simple: what is important is the quality of the drug, not its route of manufacture. Of course, part of the quality specification must include freedom from infectious agents and this includes nucleic acid.

With drugs applied to domestic animals the situation is a little different. If the animal is to be used ultimately as a source of human food, then the drug must not contaminate the slaughtered animal. This poses an interesting question in the case of animal hormones. To what extent are low levels of hormone detected at slaughter due to endogenous synthesis rather than administration? The answer may well depend on the sensitivity of the test employed; if no hormone is detected there is no problem! However, this is of little comfort to those manufacturers who have staked future profits on bovine somatotropin (BST). This hormone when administered to cows promotes a 10–25% increase in milk production (Table 14.1). Scientists in the US Food and Drug Administration, after reviewing the scientific literature and evaluating studies conducted by the producing companies, have concluded that the use of recombinant BST in dairy cattle presents no increased health risk to consumers. Nevertheless, in Europe and the USA there has been widespread opposition to its use. There are many interest groups at work: farmers, regulatory agencies, public-interest groups and the manufacturers. Consequently it is not surprising that current proposals about how to handle BST range from its unfettered use in the market-place to an outright ban for a variety of reasons, some of which are based in science and some on pure emotion.

A potential regulatory issue which has received little public debate is the use of recombinant micro-organisms to produce food. For example, to facilitate cheese manufacture strains of lactic acid bacteria which are resistant to antibiotic residues in milk and to phage attack, or which stably maintain their protease and phosphogalactosidase genes (see p. 78) could soon be in production. Will the use of such strains be acceptable to regulatory agencies? Also, fungal proteases have long been used as substitutes for calf rennet (chymosin); now that chymosin produced in *E. coli* has been found to produce gastronomically acceptable cheese will its commercial use be approved?

Even more uncertain is the regulatory fate of transgenic plants and animals intended for human consumption. Most scientists believe that knowing the function of a transgene prompts the correct safety questions but others take a more stringent view. Why there is such concern is not clear. Every day we eat the entire genetic contents of animals, plants and bacteria and we also consume the genes of incidental contaminants associated with food products. Admittedly latent toxin genes might be expressed in transgenic plants and animals but this could be assessed rapidly and simply.

Given the considerations outlined above, how does one assess the current US controversy surrounding a live vaccine against rabies? The new vaccine is a version of the vaccinia virus, which has long been used to prevent smallpox, but which has been engineered to express a protein from rabies virus (see p. 124). The vaccine has been extensively tested in animals, all of which develop immunity to rabies. The vaccine has also been tested in chimpanzees as a way of estimating its safety in humans and no adverse effects were seen in the animals or the animal handlers. More important, the vaccine has been extensively field-tested in Belgium, France, Germany and Switzerland. In Belgium alone, over 25 000 samples of bait laced with the recombinant virus were distributed over 2500 square kilometres. No abnormalities in morbidity and mortality were seen in foxes and other wild animals. Since tests began in 1987 there has been a steady decline in the reported incidence of rabies in both domestic and wild animals in the test area. Despite all this evidence of safety of the rabies vaccine it was late 1990 before approval was given for a limited field test in the USA.

Gene therapy [17, 18]

In the USA, all protocols for experiments on human gene therapy have to be approved by three groups—the investigators' local institutional review board and institutional biosafety committee, the NIH (National Institutes of Health) Working Group on Gene Therapy, and the full Recombinant DNA Advisory Committee—prior to submission to the directors of the NIH for final review. At all federal stages the review process will take place in the open. Not only will NIH committee meetings be open, but a precis of the protocol itself will be published in the Federal Register for public comment.

In early 1985 the NIH Working Group on Gene Therapy published draft guidelines in a 'Points to consider' document relating to the preparation of experimental protocols for human studies. On the basis of public comment this draft has been revised but what has been interesting is the *lack* of public comment. One interesting point did emerge. The guide-lines apply explicitly to 'recombinant DNA and DNA derived therefrom'. They do not apply to recombinant RNA, which, technically, is outside the charter of the Recombinant DNA Advisory Committee. Since RNA viruses (retroviruses) are likely to be the vectors used to transfer genetic material into the cells being returned to the human body this presents and interesting legal situation!

Three aspects of the modified 'Points to consider' document deserve comment. First, only experiments involving somatic cell therapy will be considered. The group will not consider germ line therapy protocols until somatic cell therapy has progressed and public discussion of the implications of germ line work has been broadened. Second, the patients will have to agree to at least a 3–5 year follow-up and to an autopsy in the event of death. Third, although investigators will be asked for results of experiments in non-human primates, results from experiments in other small mammals, e.g. mice or dogs, may be considered sufficient. There is no *prima facie* evidence that primate studies will provide information that cannot be obtained from other species.

What are the ethical considerations relating to gene therapy? Essentially, all observers have stated that somatic cell gene therapy for a patient suffering from a serious genetic disorder would be ethically acceptable if carried out under the same strict criteria that cover other new and experimental medical procedures. There are three general requirements. It will be essential to show in animals that:

1 the new gene can be put into the correct target cells and will remain there long enough to be effective;
2 the new gene will be expressed in the cells at an appropriate level;
3 the new gene will not harm the cell or the intact animal.

These criteria are similar to those required prior to the use of any new drug: the treatment should get to the area of disease, correct it and do more good than harm. Some flexibility may be required for some disease states, e.g. the young child with ADA deficiency for which the prognosis is certain death. Although it would be unrealistic to expect a complete cure from initial attempts at gene therapy, there must be good animal data which indicate that some amelioration of the biochemical defect is likely. Even then it will be necessary to weigh the potential risks to the patient, including the possibility of producing a pathological virus or a malignancy, against the anticipated benefits to be gained from a functional gene. For each patient there will need to be a risk-to-benefit determination but this is a standard procedure for all clinical research protocols.

A variant of somatic cell gene therapy is enhancement genetic engineering. This is not therapy of a genetic disorder but the insertion of an additional gene which produces change in some characteristic that the recipient individual desires. An example would be an additional gene for human growth hormone to produce the large size favoured by US basketball and football players. Ethically, this should not be done. Not only is there no medical need, but the short- or long-term consequences for the individual cannot be predicted. The risks are great and the real benefit nil.

The ethical and legal issues surrounding somatic cell gene therapy are really no different from those relating to heart transplant surgery. This is not the case with germ line therapy and before such therapy can be considered we will need to determine if the transmitted gene itself, or any side-effects caused by its presence, adversely affect the immediate offspring or their descendants. Satisfactory evidence that there are no long-term effects will require animal studies involving the study of several generations of progeny. The critical ethical question is whether we should ever provide a treatment which produces an inherited change and could perpetuate in future generations any mistake or unanticipated problems.

Three criteria will need to be satisfied to justify germ cell gene therapy. First, there should be considerable previous experience with somatic cell gene therapy which clearly establishes its effectiveness and safety. Second, there should be adequate animal studies that establish the reproducibility, reliability and safety of germ cell gene therapy using the same vectors and procedures that would be used in humans. It should be borne in mind that, as currently practised with transgenic animals, gene therapy does not remove or destroy the defective gene in the recipient; rather, it adds a normal gene to the total DNA complement. Thus, it will be essential to know that the new DNA can be inserted exactly as predicted, that it does not destroy the functioning of another essential gene and that it is expressed in the correct tissues at the correct time. Finally, since the societal gene pool will be affected by germ cell gene therapy the public will need to have a thorough understanding of this form of treatment and give its approval. Ethical issues aside, germ cell gene therapy could raise legal issues more complex

than those currently surrounding the long-term teratological effects of exposure to nuclear tests in the South Pacific and the defoliant Agent Orange in Vietnam.

Whenever gene therapy is discussed publicly there is a tendency for the popular press to raise the spectre of the creation of 'superhumans'. Fortunately, such traits as intelligence and fertility are polygenic and their manipulation by genetic engineering will not be possible for a long time. Ethically, most of us find the prospect of eugenic genetic engineering abhorrent and the reason we do so has been succinctly expressed by W.F. Anderson: 'Our knowledge of how the human body works is still elementary. Our understanding of how the mind, both conscious and subconscious, functions is even more rudimentary. The genetic basis for instinctual behaviour is largely unknown. Our disagreements about what constitutes "humanhood" are notorious. And our insight into what, and to what extent, genetic components might play a role in what we comprehend as our "spiritual" side is almost non-existent. We simply should not meddle in areas where we are so ignorant.'

Biological warfare

The construction of harmful biological agents, deliberately or otherwise, has always been acknowledged as the most extreme hazard associated with recombinant DNA technology. The evidence to date suggests that no new infectious agent has been created accidentally and it is difficult to perceive of anything more virulent than the naturally occurring Marburg agent or Lassa fever virus. Nevertheless, the possibility of a military use for gene manipulation has been of concern to many. It must be realized that, however created, infective agents are not good warfare weapons since the results which will be obtained with them cannot be predicted with any degree of certainty. Morbidity and mortality from communicable disease agents cannot be accurately forecast because of differences in population groups with respect to physiological, genetic and sociocultural variables, nutritional status, previous exposure to infectious agents, immunization histories and various other factors. The senior military men amongst the superpowers are well aware of these limitations. Use of biological weapons is more likely to be sanctioned by a Third World country, where the prospects for the creation of a practically useful recombinant biological weapon are unlikely.

Despite the suggested military futility of biological

weapons, does research into their creation still continue? Approximately half the nations of the world are signatories to a treaty resulting from a Biological Weapons Convention of 1972. As a signatory, a nation pledges never to produce 'microbial or other biological agents, or toxins, whatever their method of production'. However, the treaty does not prohibit studies on defences against chemical and biological weapons. This, in turn, does not prohibit the creation of 'new' organisms by recombinant DNA technology so that preventive measures can be developed. Given the current state of biotechnology, as outlined in this book, it would not be difficult to turn this research round in only a few days so that biological weapons were being produced. So is the spirit of the Convention being breached?

Biotechnology and the Third World

Many political leaders of developing nations believe that great personal prestige accrues to them if their government, or better still, a foreign company or government finances a high-technology venture in their country. The fact that the country might not need such a facility or have the infrastructure to maintain it is irrelevant. Modern biotechnology has not escaped these political fancies. This is unfortunate. Biotechnology has much to offer these countries but it is the kind of biotechnology which does not have much glamour attached to it. Around the world some 15 million young children die annually from the synergistic effects of malnutrition and infectious disease. A very large proportion of child deaths occur from diseases for which there are suitable vaccines. Two and a half million deaths are attributable to just three diseases: tetanus, measles and whooping cough. There is some cause for hope, for in the last 5 years there have been major campaigns in a number of countries including Colombia, Turkey, Bolivia, Haiti and El Salvador. In the latter country the two sides in an ongoing civil war temporarily suspended hostilities to enable mass immunization of children to take place! Diarrhoeal diseases are a major cause of death and the aetiology is frequently not known, making vaccine selection more difficult. However, the availability of clean water would reduce the incidence of the disease and would permit such simple remedies as oral rehydration therapy. Biotechnology can facilitate the supply of clean water but the methodology is not new.

The above arguments aside, are there any aspects of the new biotechnology which have particular relevance for developing countries? The answer is yes. In many of these countries the major part of the diet is derived from plants. Indeed, often a single crop is the staple foodstuff and this results in malnutrition since the plant storage proteins are almost invariably deficient in one or more amino acids. With modern plant breeding methods it should be possible to develop plant varieties with enhanced levels of more nutritious storage protein as well as ones with enhanced disease resistance. The real attraction is that these plants can be genetically engineered in a laboratory anywhere in the world. Once created they can be moved to a low-technology environment for they can be propagated by unskilled workers using traditional agricultural methods.

Further reading

GENERAL REFERENCE

Walgate R. (1990) *Miracle or Menace? Biotechnology and the Third World*. Panos Institute, London. 199 pp.

CITED REFERENCES

1 Adler R.G. (1984) Biotechnology as an intellectual property. *Science* **224**, 357–63.
2 Ratner M. (1990) T-PA trials, tribulations, and litigation. *Bio/technology* **8**, 385–6.
3 Neufeld P.J. & Colman N. (1990) When science takes the witness stand. *Scientific American* **262**, 18–25.
4 Knight P. (1990) Biosleuthing with DNA identification. *Bio/technology* **8**, 505–8.
5 Wolber P. & Warren G. (1989) Bacterial ice-nucleation proteins. *Trends in Biochemical Sciences* **14**, 179–83.
6 Lindow S.E., Panopoulos N.J. & McFarland B.L. (1989) Genetic engineering of bacteria from managed and natural habitats. *Science* **244**, 1300–7.
7 Sussman M., Collins C.H., Skinner F.A. & Stewart-Tull D.E. (1988) *The Release of Genetically Engineered Microorganisms*. Academic Press, London.
8 Steffan R.J. & Atlas R.M. (1988) DNA amplification to enhance detection of genetically engineered bacteria in environmental samples. *Applied and Environmental Microbiology* **54**, 2185.
9 Pickup R.W. & Saunders J.R. (1990) Detection of genetically engineered traits among bacteria in the environment. *Trends in Biotechnology* **8**, 329–35.
10 Collins C.H. (1990) Safety in industrial microbiology and biotechnology: UK and European classifications of microorganisms and laboratories. *Trends in Biotechnology* **8**, 345–8.
11 Ratner M. (1990) Survey and opinions: barriers to field-testing genetically modified organisms. *Bio/technology* **8**, 196–8.

12 Keeler K.H. (1989) Can genetically engineered crops become weeds?. *Bio/technology* **7**, 1134—9.

13 Hodgson J. (1989) No deliberate release in Europe? *Trends in Biotechnology* **7**, 107—8.

14 Juskevich J.C. & Guyer C.G. (1990) Bovine growth hormone: human food safety evaluation. *Science* **249**, 875—84.

15 Berkowitz D.B. (1990) The food safety of transgenic animals. *Bio/technology* **8**, 819—25.

16 Goldburg R.J. & Tjaden G. (1990) Are B.T.K. plants really safe to eat? *Bio/technology* **8**, 1011—15

17 Anderson W.F. (1985) Human gene therapy, scientific and ethical considerations. *Journal of Medicine and Philosophy* **10**, 275—91.

18 Wilson J.M., Ping A.J., Krauss J.C., Mayo-Bond L., Rogers C.E., Anderson D.C. & Todd R.F. (1990) Correction of CD18-deficient lymphocytes by retrovirus-mediated gene transfer. *Science* **248**, 1413—16.

19 Geisow M. (1991) The proof of the cloning is in the eating. *Trends in Biotechnology* **9**, 5—7.

Index